# RAT ISLAND

BY THE SAME AUTHOR

*Where the Wild Things Were*

# RAT ISLAND

*Predators in Paradise and the
World's Greatest Wildlife Rescue*

## William Stolzenburg

BLOOMSBURY

New York  Berlin  London  Sydney

Published by Bloomsbury USA, New York

All papers used by Bloomsbury USA are natural, recyclable products made from wood grown in well-managed forests. The manufacturing processes conform to the environmental regulations of the country of origin.

LIBRARY OF CONGRESS CATALOGING-IN-PUBLICATION DATA

Stolzenburg, William.
Rat island : predators in paradise and the world's greatest wildlife rescue / William Stolzenburg.—1st ed.
p. cm.
ISBN: 978-1-60819-103-1 (hardback)
1. Wildlife rescue—Alaska—Rat Islands.   2. Predation (Biology)—Alaska—Rat Islands.   3. Predatory animals—Alaska—Rat Islands.
4. Wildlife conservation—Alaska—Rat Islands.   5. Endangered species—Alaska—Rat Islands.   6. Rat Islands (Alaska)   I. Title.
QL83.2.S76 2011
333.95'1609142—dc22
2010051457

First U.S. Edition 2011

1   3   5   7   9   10   8   6   4   2

Typeset by Westchester Book Group
Printed in the U.S.A. by Quad/Graphics, Fairfield, Pennsylvania

For Don Merton and Richard Henry Kakapo

# CONTENTS

*Prologue*

# KISKA, KAKAPOS, AND
# A NOTE ABOUT WAR

A MASSIVE WILDLIFE rescue is under way, a rescue that may rank as the most promising ever waged in defense of so many creatures on the brink of oblivion.

If such rosy hyperbole reads a bit dubious in these, the dark ages of nature, the doubts come well grounded. Living species are now vanishing at an unprecedented pace in the sixth mass extinction in Earth's history—tens of thousands disappearing each year, with the body counts rising. So we're repeatedly told. We have been reminded to the point of numbness about the causes, about our clear-cutting of rain forests and bleaching of coral reefs, our melting of the ice caps and overfishing of the seas, our slaughter of the last whales and orangutans, tigers and elephants, wherever convenience or a quick profit can be had. These are the dreary tales that dominate the endangered species beat of environmental media. The victims are the poster children of conservation campaigns, and their prognoses smell of doom.

Theirs is not the story that follows. The imperiled ones covered here constitute a less familiar cast of creatures dying far from the front lines. Rarely heard is the fact that over the last three thousand years most of the planet's recorded casualties have taken place offshore, on oceanic islands. Islands have earned the ironic distinction as the most fertile crucibles and most fatal pitfalls of evolution. They have produced 20 percent of Earth's terrestrial animal species on just 5 percent of its landmass. They have also shouldered most of its extinctions, as many as two of every three missing birds and reptiles.

The survivors are still on the run. Nearly half of the species now populating world rosters of the critically endangered are island species. Their biggest threat comes in the form of animals introduced from the mainland—rats and cats and weasels, goats and pigs and rabbits, mongooses, snakes, and even ants—predators of defenseless prey, destroyers of fragile habitats, ferried to the farthest reaches of the oceanic archipelago during the human settlement of the globe.

But the tale that begins as just one more dispiriting reminder of the age of loss we live in contains a hopeful twist: An ambitious cadre of conservationists is out there now, rapidly amassing a lopsided record for rescuing these imperiled islanders. And its signature technique is astonishingly quick and thorough (if also, to some minds, brutally so). It is a technique that entails slaughtering the enemy wholesale.

Which brings up that note about war. This is a story of people who kill so that others might live. It is a theme that naturally lends itself to the rich vernacular of armed conflict. In this war for wildlife there will be combat of sorts, replete with battles and assaults, victories and defeats, bombing raids, blitzkriegs, and lines in the sand. There will be troops and battalions, SWAT teams and snipers, friendly fire and collateral casualties.

Many among those now waging this war will appreciate neither the anthropomorphisms nor the militaristic lingo, beginning most pointedly with the word "war." We are not fighting an enemy, they will say. These invaders did not come with intentions of doing harm, and we would just as soon not do them any harm in return. But time is short, extinction is forever, and there is just no other way, they will say.

Fair enough—in a more perfect world. As it stands, the warriors come armed with their own emotional terminology. Alien, plague, invader—these are tags of the conservation community's own choosing, affixed to the creatures they are compelled to kill. Charles Elton, the man who more than fifty years ago wrote the bible of invasion biology, called them "ecological explosions."

So with combat clichés aforethought, let the battles begin. Let them begin, say, on a remote and rawboned island called Kiska, eleven hundred miles west of mainland Alaska, where it turns out that a few authentic bombs—of the more conventional, high-explosive, antipersonnel type—do come into play. During a pivotal episode in the North Pacific theater of World War II, Kiska came under regular poundings from American bombers and battleships, which on busier days rained more than half a million pounds of explosives, with intentions of routing and eradicating the invading Japanese enemy.

The relevance of that war story to this war story involves yet another invasion, under way and underfoot even as the artillery shook the hills. It seems that sometime during the commotion, under the clanking din of anchor chains and the droning of diesel engines, emerging from the hold of a battleship or a cargo container, a rat or two secretly accompanied the soldiers ashore to Kiska.

Some half century later this unheralded little footnote of biogeography would come to weigh as heavily on Kiska's destiny as any human intrusion ever has. It was around that time that the island's indomitable little castaways finally made their way over a dozen miles of tundra, past a four-thousand-foot, ice-encrusted volcano, to a tremendous field of black lava boulders sloping into the sea. There the rats discovered a gathering of birds, otherwise known to a few privileged people as one of nature's great spectacles.

The celebrated performance on Kiska commences with the long northern twilight of late June. It begins offshore, as a chain of living clouds rising over the gray horizons of the Bering Sea. The clouds billow on approach, morphing into fantastic forms. A sphere becomes a scythe, a serpent, a genie emerging from a lamp.

The masses have been variously likened to swarms of bees, plagues of locusts, herds of bison on a prairie sea. Boat pilots have mistaken one for a wall of water coming to swallow the vessel. The swarms are in fact composed of many hundreds of thousands, perhaps millions, of what on final approach become flocking birds.

The flocks combine two varieties of a rather diminutive sea-bird. Two out of ten in the crowd tend to be crested auklets, each jacketed in black over gray, with a rakish plume arching from the bridge of its beak. The remainder and bulk of the multitude are least auklets, a smaller, less gaily ornamented congener of the crested species, in the general life-form of a miniature, stub-billed penguin. Auklets are able fliers of two mediums, their wings propelling them underwater like submarine falcons in skilled pursuit of tiny planktonic animals, and flying them aloft in those dramatic aerial displays for which Kiska has become legend.

Against the softly glowing backdrop of Kiska's snowcapped volcano, the incoming flocks swirl. The auklets' flight is erratic yet precise, their split-second dodging and feinting synchronized en masse as if by the single mind of a superorganism. A tendril of thousands splits off and ascends the slope, climbing, climbing, all but disappearing into the distant heights, to finally turn as one and plummet seaward again, ripping the air with the roar of jet fighters.

It is beneath the lava boulders that their nests and progeny are waiting, but the auklets resist. They come sweeping landward, squall after squall of birds raining against the rocky slopes, only to break off and circle for another surge. Behaviorists explain these false approaches as antipredator maneuvers. Peregrine falcons and bald eagles patrol the colony's airspaces, picking their moment to dive and snatch a conspicuous outlier from the masses. Glaucous-winged gulls lurk about the boulders to surprise the unwary auklet at its door. These are the high-stakes dances of predator and prey, choreographed over the eons since the birds first took wing. Only after all precautions and security protocols have been satisfied do the skittish swarms put down on the roof of their rock fortress, to preen and chatter before vanishing below into the maze of a million nests. For all the drama, a few auklets die, a few gulls and falcons feed, and the balance is overwhelmingly displayed in the auklets' favor, in the colossal summer flights that fill the skies above Kiska. Or at least that's the way it used to work.

Cue again the rats. The brown rat is native to the mainland of Southeast Asia; the least auklet to a few dozen islands of the Bering Sea. The two have never as a pair learned such a dance. The rat, upon finding an auklet on a nest, bites a hole through the back of its head, eats its brain and eyeballs, then stashes the rest. And repeats, the frequency depending on how many auklets it happens upon, a number that in the subterranean metropolis of Kiska rapidly mounts. Habit overriding hunger, a single rat has been known to gather one hundred fifty auklet bodies to its cupboard, most of them largely intact and destined to rot. The auklet, for its part, apparently sits confused through the slaughter. Its memories of such dangers were left behind in a distant past. Such four-legged predators were among the chief reasons that the auklet in its evolutionary beginnings abandoned the dangerous continents for the relative seclusion and safety of the islands.

Which is to say that with rats having finally found them, the future of least auklets on Kiska, and of the whole mighty spectacle they have come to be, now lies in question. Which raises a few others: Should they, need they, can they be saved?

As of the summer of 2010, conservation specialists had conducted more than eight hundred eradications of destructive mammals from islands they had breached with human help (most of them coming with quickening pace in the last twenty years). The eradicators have covered islands across both hemispheres, from tiny tropical atolls in the sunny Pacific to howling wildernesses of snow and tundra in the high latitudes. Among them all, Kiska remains a singular prize. Those who would dare to defeat the rats of Kiska see a potential payoff amounting to millions of living birds rescued with one swipe, and the satisfaction of securing a living wonder of the world.

Those who would so dare have a few sizable obstacles in their path. The island is staggeringly huge, at more than a hundred square miles; hellishly stormy on all but a few rare days of the year; and eleven hundred miles from almost anywhere. It is lorded over by a snowcapped, fog-enshrouded volcano that still features the tail wings of errant warplanes poking out of its flanks. It also harbors great numbers of bald eagles and other birds very likely to die in the crossfire of what would most certainly entail an aerial assault with poison. As irresistibly tempting as they come, the most daunting battle yet in the war for the world's islands is the one to be waged for Kiska. If it's not the one to be waged for the kakapo.

The kakapo is an enormous, moss green parrot from New Zealand, with a life history almost perfectly contradicting the least auklet's. The kakapo does not swim, nor does it fly. The kakapo, at a hefty six to nine pounds, is the largest and least airworthy of all parrots; the least auklet, at five ounces, is the smallest of its clan of auks. The kakapo is one of the most painstakingly probed and pampered animals on the planet, every member of its species held under twenty-four-hour surveillance on two New Zealand islands groomed especially for the birds' comfort and safety, every moment of need met with offerings of food and shelter, medical care, or even assistance at mating. The far-lesser-known least auklet passes its summer tending subterranean nests typically far out of human reach; the remainder of its year is spent on the water in some far corner of the stormy sea where nobody has yet found it. The world population of the least auklet is wildly estimated at ten million, making it the most numerous seabird in the northern hemisphere; that of the kakapo is known to the single digit—the count at the moment, 122.

The two do share one critical piece of common ground. The kakapo, like the auklet, evolved in a world devoid of land-bound mammals. The greatest danger it typically faced, as a chunky walking parrot in its primeval New Zealand, was a limited suite of native raptors hunting from above. The kakapo's answer was its shrub green camouflage, a nighttime schedule of activity, and, as a last resort in the face of danger, a habit of freezing in its tracks.

Such cryptic defense eventually came to approximate suicide in a land invaded by terrestrial predators, particularly those with noses. (The kakapo exudes a scent variously described as that of honey, or freesia flowers, or perhaps like the inside of a clarinet case—if you've ever been there.) The flightless kakapo typically nests on the ground, in holes or tree pits readily located and looted by cat, dog, or weasel. The mother kakapo with egg or chick gets no help from the father; she wanders all night foraging for food, leaving her offspring unguarded for the taking, and herself vulnerable to attack on the trail at a time when foreign predators are most likely to be about. Her eggs are small for such a large bird, small enough for rats or maybe even mice to eat them. The point is, it would be challenging for even the most sadistic of bird gods to construct a specimen more flagrantly begging to be slaughtered by terrestrial carnivores than the kakapo.

Which is why, over the course of eight hundred years and a procession of invading humans, cats, dogs, weasels, and rats, the kakapo's status has fallen from the ranks of the most ubiquitous birds in the country to that of a precarious castaway huddled on two tiny makeshift homes under high security.

The kakapo and the least auklet, falling prey to a dangerous new world of foreign predators, cover the extremes of the world-wide rescue now under way. To free the auklets of Kiska will entail an assault harking back to World War II. To restore the kakapo's rightful home will require retaking great swaths of New Zealand. There are certainly more than a few who would consider such campaigns as hopeless, who see the continuing extinction of the masses as inevitable. But as mentioned above, theirs is not the story that follows.

# Chapter 1

## OVER THE BLUE HORIZON

*Let it be remembered how powerful the influence of a single introduced tree or mammal has been shown to be.*
—CHARLES DARWIN

SEVEN CENTURIES AGO, from a tropical beach in the South Pacific, a boat set sail. From Tahiti or Rarotonga, Tubuai or Rangiroa, the precise port of departure long since lost in the haze of cultural memory, the clan of Polynesian seagoers sliced into the surf on a double-hulled canoe, a catamaran buoyed by two great hollowed trees. Its crew steered south by southwest, into unexplored waters.

Why they set sail remains a question for the ages. Their leader may have been a young man with political ambitions, whose only hope for becoming island chief was to find an island of his own. They may have been outcasts, forced seaward by crowding or banned by society. Perhaps they were simply explorers, heeding the human itch to know the other side of the horizon.

Moved by whatever push or pull, into the blue unknown they went. They steered by the stars, by Alpha Centauri and

Beta Centauri, by the familiar beacons and bearings of distant galaxies. By day they watched for signs of land, in the passings of coconuts and driftwood, the swim-by of a sea turtle. They marked subtle changes in the behavior of waves and swells, reading currents for the particular curves imparted by intervening shores. They scanned the skies for clouds signaling the billowing of air over sun-heated hills, or the purposeful flight of a seabird, suggesting a nest not too far away.

Several weeks and three thousand miles into the void, the Stone Age seafarers at last sighted shore. Behind it rose the tall green jungles and mountain gorges and rushing rivers of a land spreading farther than any they had ever known. Aotearoa, they would come to call it—land of the long white cloud. As they unpacked their stores of taro and sweet potato, their fishing hooks and axes, one among them retrieved a length of hollowed log, an elegant vessel carved in the likeness of a canoe and capped at the ends. It was handled only by its *tohunga*, its expert, with a purpose deserving of precious cargo, and was soon to be carried into the forest, accompanied by prayer. At the proper place, the *tohunga* ceremoniously lowered the vessel and opened its latch. And onto the ground and into the forest scurried a family of rats. *Kiore*, the rats were called.

## Pacific Boneyards

The Polynesians' landing of Aotearoa, the last great unpeopled landmass of Oceania, was in a grander sense the end of a far longer voyage that had begun three thousand years before, when their ancestors set sail from somewhere in the Bismarck Archipelago, north of New Guinea. Beginning as small forays between islands beckoning from the horizon, the trips grew longer, penetrating tens and hundreds of miles eastward into the Pacific.

Onward the people sailed and settled. On a New Caledonia beach named Lapita, they left a buried cache of pottery, of a signature design that twentieth-century archaeologists would later unearth like calling cards across Melanesia. The archaeologists would name the Pacific pioneers the Lapita.

The Lapita set up seaside villages, sheltered in thatched-roof houses raised on stilts. They cleared forests and cultivated crops, hunted and gathered wild foods from forest to shore to sea, and fired their elegant pottery. And inevitably, either with a shove from society or a romantic pull of the beckoning horizon, another band of explorers would sail eastward into the unknown. By 1000 B.C. they had pushed nearly four thousand miles, to the shores of Tonga and Samoa. By 600 A.D. the Lapita's Polynesian descendants had reached Hawaii, with the coast of South America soon on the horizon. By the time they made their last great push, south by southwest against the prevailing trade winds to Aotearoa, there was hardly a mote of land in the South Pacific that they had not either settled or inspected for its livability.

In the mid-1700s, Captain James Cook and the brigade of European explorers who followed him began what would amount to their rediscovery of the Pacific. They were to find in this island-peppered expanse of sea, hundreds and thousands of miles adrift from anywhere, people and languages and ways of life remarkably mirroring one another.

With the European ships came the familiar pattern of missionaries, traders and exploiters, imported diseases, subjugation, and slavery. Societies collapsed, forests fell, birds and seals and whales vanished. Cook's arrival was the beginning of Eden's end. Or so the story once went.

In 1971, on the Hawaiian island of Molokai, a naturalist named Joan Aidem noticed in her wanderings something odd poking out of a windblown sand dune. As Aidem brushed the sand aside, the oddity became the more obvious form of a bone. Beneath it lay the entire skeleton of a bird—a big bird, apparently some odd sort of waterfowl. Aidem sent the bones off for identification, and they eventually wound up in Washington, D.C., in the Smithsonian Institution's avian collections. The skeleton aroused more than a little curiosity. Its breastbone was shrunken; its wings were withered. It was flightless. And it was huge. It was given the Latin name *Thambetochen chauliodous,* meaning "monster goose." The live version of the bird had never been seen by anybody in modern history. Cook and his men had never mentioned it, nor had anyone from the parade of explorers and naturalists who had followed over the next three centuries.

The monster's bones opened a crack of light into the Hawaiian Islands' prehistory. Hawaii until that time had been a notoriously bleak prospect for modern hunters of ancient birds. Its volcanic paroxysms and grindings had stacked precarious odds against the preservation of brittle bones. Aidem's find aroused fresh curiosities, particularly for the Smithsonian's chief avian paleontologist. Storrs Olson wondered how many more Hawaiian oddities lay hidden. Five years later, with grant money finally in hand, he set out in search of Hawaii's unlikely fossils. Soon joined by his protégé Helen James, the two embarked on what would prove to be a history-remaking odyssey among Hawaii's missing fauna.

Olson and James scouted the terrain and geological formations, following tips from the local experts. And from the supposed desert of paleontology sprang oases of bones. In old lake beds and ancient dunes, limestone sinkholes and lava tubes, caves and cliffs, the two started gathering. And as they labored, Hawaii's list of birds began to bulge.

From the bits and pieces of bone appeared geese and rails, and a small seabird of the petrel family. On Kauai, Olson and

James pieced together the bones of three more species of goose, for all purposes flightless. They assembled another rail, a long-legged bird that had once crept among the reeds of Hawaiian marshes. It too was flightless. They found a long-legged owl, and a host of finchlike songbirds, missing members of Hawaii's modern array of honeycreepers. On Oahu, more of the same: several geese, a couple of rails, a long-legged owl, more honey-creepers, plus an eagle and a hawk. There emerged two species of extinct crow. On Molokai the pattern repeated: flightless geese, an eagle, a hawk, a tiny flightless rail, another long-legged owl, another crow, more honeycreepers. They reconstructed an odd sort of ibis, a stubby, sturdy-legged, flightless skulker of for-est floors, far removed from the twiggy, stilt-legged wading bird of modern form. Maui: more flightless ibis, more flightless geese, more rails, another owl, another honeycreeper. When all were tallied, Olson and James had uncovered thirty-nine species of Hawaiian birds never seen by modern ornithologists.

The birds appeared in outlandish forms, an intriguing lot of them evolutionarily reconfigured for life on foot. It was a para-doxical kingdom of grounded birds, walking the forests, filling the niches of deer and squirrel, tortoise and hare, reinventing the world of land-bound quadrupeds left far behind on the main-land. It was, in essence, an avifauna of feathered mammals.

As intriguing as the birds' unveiling were the accompanying artifacts with which they were commonly buried. Often the birds lay among ancient hearths and pits and grindstones, in soils marbled with charcoal. They lay in middens heaped with the shells of mussels, the bones of fish and chickens.

These birds, unknown in the annals of ornithological his-tory, had lived in a time of people. The biota that Captain Cook and his eighteenth-century successors had cataloged on Hawaii had embodied a mere skeleton crew compared with the magnificent menagerie that had met the islands' pioneers. Sometime in those eleven hundred years between the landing of the Polynesian people and the coming of Westerners, Hawaii's avifauna had been pared by half.

It was a bittersweet glimpse into an era of evolutionary oddities that the modern world had only barely missed. It flagged a disturbingly repetitive pattern of human arrival soon followed by waves of island extinctions. And it raised the questions of what else might have lived out there, in the far-flung expanses of the island universe, and how precisely they had died. They were questions soon to hijack the career of a young grad student new to Olson's Smithsonian lab.

With the prevailing wisdom of Hawaii's pre-European purity now trashed amid the native Hawaiians' bone piles, paleontologist David Steadman cast his suspicions over the entire breadth of the South Pacific. He began digging back through the three millennia and three thousand miles of the Lapita's ocean-conquering odyssey. And under stronger light, Hawaii's explosive feat of speciation and meteoric crash of extinction appeared far more pedestrian.

Steadman's reconstruction began with the rails. So many of the islands across the Pacific had developed their own brand of flightless rail, that clans of chickenlike marsh birds that had apparently made an art of colonizing the Pacific. The textbook rail naturally tended toward smallish wings and a preference for skulking and hiding over flying from danger. But once airborne with the seasonal beckoning of migration, the little-winged birds became aeronautical demons. Whether under their own heroic powers or the hijacking winds of a great storm blowing them to hell or Tahiti, the rails in time conquered the breadths of Oceania.

Once landed on a remote island, the rail would find itself, for all practical purposes, stranded in paradise. Life in the predator-free kingdom and the selective pressures of evolution would fast begin reshaping. No need for flight, no need for big breastbones or long wings or the energy-eating muscles to power them. Less energy invested in flight meant more energy to build big bills for eating, big guts for digesting, and big legs for getting around. The leap from airborne to earthbound was actually rather easy. The plump body, little wings, big legs—it was as if the island bird had flipped a genetic switch, forgoing its adult form for that of an oversize chick. And in a blink of geological time, winged bird thus became walking bird.

And so it happened time and again, across the Pacific. Each of the nineteen tropical Pacific islands Steadman studied told of a recent past that had birthed and harbored up to four unique species of flightless rails. Extrapolating his sample to the whole of Oceania, Steadman figured an island roster of rails amounting to some two thousand species.

Most, however, were now gone, vanishing in step with their island's occupation by canoe-sailing humans. And the mass exodus was far more than a rail phenomenon. As Steadman plumbed the fossil beds of the Polynesian heartland, the ranks of the missing multiplied. In the Marquesas, eight of twenty species of seabirds were gone, the survivors banished to tiny offshore islets. On Ua Huka, five of six species of pigeons and doves, three species of rails and parrots—gone. Five hundred miles east of Fiji, in the sea-cliff caves of the island Eua, Steadman extracted the bones of thirty-three species of land birds and seabirds no longer to be found.

The greater wonder to Steadman was that any birds remained at all. "Extinction is what we have come to expect on islands," he wrote, "survival is the exception." On a planet whose average rate of extinction amounted to one missing species of life every million years, this was an episode of nearly instantaneous implosion.

In 1995, Steadman entered a major note in the scientific literature, publishing in the journal *Science* what would become a classic paper. It was at once an announcement of discovery and a death toll. When all was tallied, Steadman conservatively estimated an average of ten species or populations having disappeared on each of Oceania's eight hundred–odd islands. Eight thousand populations had disappeared. Some of them had managed to survive elsewhere, in little island hideouts here and there, but a terrific number had represented the last of their kind. More than two thousand species had been swept from Oceania before the Europeans' supposed apocalypse.

There had been major extinctions in human time before, all of them far better publicized than this one. The most famous of them had been made so by Steadman's own mentor Paul S. Martin. Thirteen thousand years ago, soon after Siberian mammoth hunters made their way across the Bering Strait, North America lost three quarters of its great mammals, marking the demise of its mammoths and camels and horses, giant bears and giant ground sloths and saber-toothed cats. Pleistocene overkill, Martin called it. Yet for all its dramatic crashing of giants, the megafaunal blitzkrieg of North America removed but a few dozen species. South America, Australia, New Zealand, and Madagascar all suffered similar mass extinctions of their megafauna. All followed close on the heels of their settlement by humans; all had damning bone piles and spear points as smoking guns; none came close to matching the sheer numbers of species Steadman was now counting among Oceania's missing. With the invasion of the Pacific islands, Earth's avian roster was pared by 20 percent. Oceania had hosted what Steadman would announce as "the largest single extinction event ever detected for vertebrates."

How those thousands of species had ultimately died amounted to what Steadman summed up as "the triple whammy." With the first scrapings of sea canoes against sandy shores came three major forces against the life of islands. The canoes landed hungry people bringing pointed weaponry and fire, colonists who of course hunted the island birds for food and feathers, and who habitually burned and cleared for their crops what had once been the birds' forests. Beyond hunting and habitat loss, the third force came in the form of an accomplice.

The Lapita and their Polynesian descendants habitually stocked their canoes with supplies, not only for the long haul across the water but also for the extended stay once they arrived. They brought their taro and yams for planting and their stone adzes and fire skills for clearing and burning the fields. They also brought animal food, in the form of the domestic chicken, pig, and dog. And most religiously of all, the Pacific voyagers packed their rat.

*Rattus exulans*, the Pacific rat, was a constant companion of the seafaring clans. On almost every one of the islands that bore any sign of Polynesians, there were signs of their rat. The rat often traveled with a purpose, as a snack for the long overseas trips and as a self-perpetuating crop of protein to be planted and harvested in the new homeland. For the twentieth-century archaeologist digging up Oceania's past, the bones of *Rattus exulans* became a marker of human habitation as sure as the shards of Polynesian pottery. A sleek little mammal from Southeast Asia—with the climbing skills of a squirrel, but with no wings or fins—the Pacific rat had a presence across thousands of miles of the world's largest ocean that could only be explained by human transport. And in time the rat would be recognized as a force of nature to nearly rival its keepers.

## No Moa

It wasn't until the thirteenth century—long after the settlement of Hawaii, of Easter Island, of nearly every speck of habitable land across the breadth of the South Pacific—that the little clan of Polynesians and the rat they called *kiore* finally set sail on that long journey southward, tacking into the trade winds, to the last great unexplored landmass of the Pacific.

The colonists of Aotearoa had landed well. For the *kiore*, there were fruits and nuts for hoarding, edible insects the size of mice, lizards and little birds with undefended eggs and nestlings. For the *kiore*'s people, the Māori, there were beaches where they could comb for mussels and crabs, club a cornered seal, or scavenge a beached whale. There were waters in which to dive for conchs, spear and hook fish, harpoon a dolphin. Seabirds by the millions nested on the cliffs and headlands, to be plucked like berries. Inland lay great forests, with rich soils for growing crops, but also, as they were soon to discover, two-legged monsters.

In their three-thousand-year tour of the Pacific islands, the Aotearoans' ancestors had walked among giant geese and rails that reached to the waist. But there had been nothing to approach the enormity of the creature now standing before the new settlers. This thing walked on massive clawed feet affixed to two bony legs the thickness of rowing paddles. Its rotund body was covered in a shaggy cloak of plumes, narrowing to a long serpentine neck and a blunt and sturdy beak. From head to toe, the beast towered as tall as any two of the clansmen. It is not recorded which of the strangers fled upon the first encounter, but very soon thereafter—as the fossil record would abundantly reveal—one definitely became the pursuer.

The giant moa, as the first people of Aotearoa came to learn, made for epic feasting. A guild of moa hunters rose to the task, learning to avoid those treacherous feet and to spear and tackle and subdue these walking bonanzas of bird meat.

The moas were of a family of some fourteen long-distant cousins to the ostrich, some dwarfing their African counterpart, some as small as a bantam hen. All were flightless, and all were rabidly assailed. The moas' eggs, laid on nests of naked ground, became giant omelets, their shells drinking cups. Bones piled high in the middens of the moa hunters. One butchering site, excavated five hundred years after the hunting had ended, contained the remains of 678 moa. The bone gardens were so thick in places that industrial-age entrepreneurs came to mine and market the refuse as fertilizer. Within perhaps a century of the moas' meeting with the first people of Aotearoa, no moas remained.

The moas of Aotearoa were to become the symbolic victims in a country of evolutionary oddities on the verge of plunder. The landmass had been born in the breakup of the supercontinent Gondwanaland some 130 million years before the human form had been conceived, coming unstitched from Antarctica and Australia and drifting away on its own course. In its early departure Aotearoa had left behind those creatures that would one day be mammals and had set sail as a country stocked with primitive insects and spiders, dinosaurs, reptiles, and birds.

And from the threadbare cast of animal designs, the eons of isolation and evolution built a bizarre menagerie. With the demise of the dinosaurs, their descendants took over. While the rest of the world underwent a mammalian revolution—a flowering of furred creatures from mice to mammoths—the mammalian niches of Aotearoa blossomed with birds.

Aotearoa was the evolutionary crucible that produced the modern kiwi, a wingless bird wearing its nostrils on the far end of a dipstick bill, for probing little creatures deep in the forest duff. About the size of a domestic chicken, the kiwi lays an egg six times the size of a chicken's egg—a quarter of the mass of its mother. Unlike flying birds, whose bones are hollow for lightness, the hefty kiwi has bones filled with marrow, like those of a mammal.

With the development of the moas, the biggest of them standing nearly twelve feet tall and weighing a quarter ton, Aotearoa had itself the ecological counterparts of the horse and the camel. No tigers or wolves existed to hunt them, but there were equivalents patrolling the skies, most spectacularly in the form of an enormous eagle, *Harpagornismoorei*, with a ten-foot wingspan and meat-hook talons that probably gave young moas their best reason for running.

Vying for the title of the oddest of the lot was the kakapo, Aotearoa's answer to the possum. The kakapo was, at its ancestral core, a parrot, beyond which comparisons became vague. It grew big and chunky, up to nine pounds heavy and two feet high, more closely resembling an owl. It was the heftiest parrot ever, a feat made possible in its abandonment of flight. The kakapo scuttled about in the understory of the brushlands, hobbitlike, foraging for fruits and nuts and leafy greens. It sometimes climbed and clambered about the trees. Its only ingrained fears came from the skies in the form of raptors, which it escaped by hunkering down and hiding by day.

The grounded existence and cryptic defense that had served the kakapo for so long would soon render it easy meat in Aotearoa's new era of terrestrial predators. As moas went scarce, and as the country's swans, geese, giant rail, and goshawks disappeared too, the moa hunters turned their sights and their dogs on the kakapo. The Māori's *kuri*, a wiry little dog that had accompanied the Lapita seafarers from their ancestral ports off New Guinea, would sniff out birds hiding in the thickets. If a kakapo lay hunkered in its burrow, a barbed stick would snag and drag the growling bird to the hands that would wring its neck.

The kakapo's flesh was a delicacy. Its soft green feathers, when woven, became the fabric of the cloaks and capes of chiefs. On feast days, partygoers wore earrings strung with the heads of kakapos.

What the hunters and their dogs and their fires didn't manage to obliterate of Aotearoa's wildlife, the *kiore* often did. When the beech or rimu trees produced a particularly good crop of seeds and nuts, the *kiore* periodically irrupted in plagues and scoured the forests top to bottom. *Kiore* ate the forest fruits that fed Aotearoa's animal kingdom. They ate the animals too. They feasted on Aotearoa's giant flightless beetles, on the eggs of Aotearoa's nocturnal lizards, gnawed through the shells of the island's giant land snails. They ate the eggs and hatchlings of Aotearoa's tuatara, Earth's sole remaining member of a lizardlike clan of reptiles that had walked with dinosaurs two hundred million years before. *Kiore* ate Aotearoa's giant weta, a cricket the size of a mouse. They ate the eggs and chicks of colonial seabirds. Six species of little songbirds disappeared with the *kiore*'s arrival. *Kiore*, it would later be suggested, as an eater of eggs and a competitor for forage, may have even helped slay the giant moa.

*Kiore* spread in advance of their hosts, multiplying exponentially, swarming over the virgin Aotearoan candyland—"a grey tide," wrote the paleontologist Richard Holdaway, "turning everything edible into rat protein as it went."

Survivors of Aotearoa's invasion retreated to tenuous safety offshore. The myriad coastal islands of the island nation became the last refuge for the New Zealand snipe, for the sitting duck called the Auckland merganser, and for the dinosaurian throwback the tuatara. But even the farthest sanctuaries would soon be too near.

## STRANGERS BEARING GIFTS

In 1642 a Dutch sailing fleet commanded by Abel Tasman sighted the Aotearoan homeland, "a large land, uplifted high." Tasman's men were likely looking upon the frosted peaks of the coastal range now called Fiordland, on the southern island of what his cartographers would later name New Zealand. A canoe loaded with islanders came out to meet Tasman and, not trusting the looks or intentions of the strangers, rammed one of his vessels and killed four of his crew. Tasman's men returned the greeting, killing several of the islanders. Welcome to Aotearoa. Without stepping ashore, Tasman beat a hasty retreat, away from the newly christened Murderer's Bay, and sailed north for what he hoped would be warmer receptions in Fiji.

A century later the outside world came knocking again, this time sticking around for keeps. In 1769, HMS *Endeavour*, commanded by James Cook, made its way from Great Britain across the Atlantic, around South America's Cape Horn, and west into the Polynesian universe of the South Pacific. On October 6, with the help of a Tahitian guide, Cook and the crew of the *Endeavour* reached the land of New Zealand and began charting its shores and meeting its residents, the Māori.

To Cook, the Māori were a paradoxical people with a gift for elegant gardens and a fearsome reputation for eating their enemies. Cook traded cloth, beads, and nails for the Māori's fish, sweet potatoes, and dog-skin cloaks. Occasionally the two traded aggressions, canoe-loads of Māori singing heartily of killing

Cook and crew, Cook and crew returning the compliments with guns and cannons.

Cook would eventually make three Pacific voyages, stopping each time in New Zealand and leaving more than trinkets and the occasional skirmish behind. His boats came increasingly loaded with animals from home. "Floating menageries" of cattle, sheep, goats, pigs, rabbits, chickens, turkeys, geese, ducks, peafowl, dogs, cats, cockroaches, and rats toured the Pacific, as guests of Captain Cook. The barnyard passengers were regularly unloaded as gifts to the islanders. The ships' stowaway vermin helped themselves ashore. Cook's mooring lines provided his ships' rats with the equivalent of gangplanks, spilling them ashore like convoys of cruise-ship tourists.

The rats of Cook's ships were an animal apart from the Māori's *kiore*. They were *Rattus norvegicus*, brown rats originally misnamed Norway rats, natives of northeast China that had mastered a rewarding vocation raiding the grain bins and garbage heaps of Western civilization, and stealing rides on its sailing ships around the world. "They stood in their holes peering at you like grandfathers in a doorway," wrote a young adventurer named Herman Melville from aboard one of the whaling ships that would one day inspire *Moby-Dick*. "Every chink and cranny swarmed with them; they did not live among you, but you among them."

Once ashore, ferried either aboard cargo or by the paddling of their own little feet, the brown rats immediately made themselves at home. Not quite as able climbers as the *kiore*, they made a better living on the ground. They combed the beaches for shellfish and sand fleas and stranded marine life. They prowled the seabird colonies that had come to such shores as New Zealand's to avoid their type. The brown rats swept inland, pushing the smaller *kiore* aside, making meat of forest birds so conveniently nesting on the ground. They spread through the mountains and forests.

By the 1870s the plague of rats had become a common entry in the journals of New Zealand's colonial naturalists. "This cosmopolitan pest swarms through every part of the country, and nothing escapes its voracity," wrote the ornithologist Walter Buller. "It is very abundant in all our woods, and the wonder rather is that any of our insessorial birds are able to bear their broods in safety. Species that nest in hollow trees, or in other situations accessible to the ravages of this little thief, are found to be decreasing, while other species whose nests are, as a rule, more favorably placed, continue to exist in undiminished numbers."

"These rats are the great enemies of birds, and any bird living or breeding near the ground has but a small chance of existing," wrote the ironic Andreas Reischek, a noted collector and plunderer of New Zealand's scarcest avifauna. "They play havoc alike with eggs and young, and even attack the parent birds . . . It took five months of shooting, poisoning and trapping before they showed signs of decreasing around camp."

By the latter half of the nineteenth century, what was left of the New Zealand fauna had been invaded by a third species of rat. *Rattus rattus*, the black rat, had made its way from its homeland in Southeast Asia with the sailors and whalers of the southern seas. More lithe and athletic than the brown rat, and a better climber of trees than the *kiore*, the black rat was the most versatile killer of the trio.

It got worse. Captain Cook and his cohorts, whose ships came infested with rats, brought cats to hunt them. It became routine at ports of call for the ship's cat to stretch its sea legs, saunter ashore for a stroll through the local woods, and make game of the island birds and lizards. It was hardly surprising that many cats, having tasted the wild life, never returned to ship. By the turn of the nineteenth century, feral cats could be found across New Zealand, from the coasts to the mountain snow lines.

So too came the rabbits. Introduced as game in 1864, the rabbits did what rabbits do, and within a decade New Zealand was boiling over with them. They mowed their way through the pastures of the vast sheep empires that had built New Zealand's new civilization. Sheep starved en masse, sheep farmers clubbed and killed rabbits by the millions, and still the rabbits kept coming, torching the countryside as they went. Rabbits might have ranked as the worst idea in the ecological history of New Zealand, if not for the ensuing harebrained scheme to rid them.

In their panic to save their sheep from their rabbits, the governing authorities of New Zealand in 1882 began shipping would-be rabbit predators. Three species of lithe, low-slung mammalian carnivores of the Mustelidae family—the ferret (or polecat), the weasel, and the stoat—were gathered up from Great Britain and turned loose on New Zealand. Frenetic, high-energy hunters at home in tight spaces, the mustelids were infamous for their unnerving mix of curiosity and giant-killing savagery. The stoat, at ten ounces, was practiced at grappling with rabbits five times its size, dispatching its prey with a penetrating bite through the back of the skull. But the little carnivore also came with brains, and the common sense to take advantage of any and all trusting songbirds and sitting ducks and grounded parrots that epitomized the New Zealand avifauna.

Buller, the high-profile bird enthusiast, voiced the naturalist's outrage over the mustelid liberations. "The legislature having rejected the proposed measure for prohibiting the introduction of polecats and other noxious animals into this colony, nothing now remains for us but to sound the note of warning before it is too late, and by directing public opinion to the subject, to mitigate the danger of our being overrun with one of the worst of predaceous vermin."

All such warnings duly ignored, the New Zealand authorities went on shipping mustelids. And as predicted, the immigrants took to stuffing their larders with New Zealand's trusting avifauna, while the rabbits went on ravaging the sheep range.

To which the sheep lobby responded with the stupendous logic of introducing more foreign predators. They rounded up cats from town, tossing them like grenades upon their rabbit-ravaged fields. And onward the rabbits and stoats and cats merrily marched. Finally, in 1939, in a darkly comical parody of closing the barn door behind the missing horse, the bumbling new colonists of New Zealand, with their country in ecological tatters, enacted a useless bounty on stoats, offering two shillings a tail.

By then, half of the native bird species of New Zealand were gone, and nearly half of the survivors were circling the drain. By the 1890s explorers and collectors heading into the glaciated peaks and valleys of Fiordland were finding the supposed wilderness already ransacked. Those accustomed to traveling light and growing fat off the land now faced starvation in deserted forests. The birds that they had once so blithely gathered with guns and dogs and sticks, birds that had once eaten out of their hands, were no longer to be found.

"The Digger with his Dogs, Cats, Rats, Ferrets and Guns has nearly exterminated the Birds in the lower reaches of the southern rivers," reported the explorer Charles Douglas, whose own crews had once pillaged their way through this virgin territory, piling up hundreds more birds than they could eat, and leaving the rest to rot. "The cry of the Kiwi is never heard and a Weka is a rarity. The Blue Duck once so green, is as carefull of himself as the Grey and the Robins are extinct."

Those few species that endured the invasion owed much to their wings and good fortune to find pockets of predator-free refuge elsewhere in their shrinking universe. While those birds stranded in hostile territory and lacking the option of flight more often suffered the ultimate demise. And none more infamously than one tiny songbird named the Stephens Island wren.

## THE LEGEND OF TIBBLES

The Stephens Island wren was a tiny flightless species that within a year of its discovery was extinguished at the paws of one lighthouse keeper's cat named Tibbles. Thus reads the popular legend of what is commonly claimed as the only known instance of a single individual driving a species to extinction. What actually befell the Stephens Island wren was a bit more involved, if ultimately no less catastrophic.

In 1892 work crews came to build a lighthouse on Stephens Island, an otherwise uninhabited square mile of wilderness in Cook Strait, at the northern tip of New Zealand's South Island. In 1894, sometime after the lighthouse began operating, a cat belonging to one of the new residents started coming home with little dead birds in its mouth. An assistant lighthouse keeper and amateur naturalist named David Lyall skinned one and sent it to the ornithologist Walter Buller. The sight of the skin excited Buller, as that of a bird "entirely distinct from anything hitherto known." The elated ornithologist wrote Lyall, "There is probably nothing so refreshing to the soul of a naturalist as the discovery of a new species."

As the news of the curious new wren of Stephens Island spread, the celebrations turned sordid. There was fame and fortune riding on the head of the unique little bird. Buller, an ardent collector and profiteer, started plying Lyall for more birds. So did Henry Travers, a natural history entrepreneur and a noted broker of such rarities. The shrewd Travers secretly talked Lyall into diverting the specimens his way, and thereafter began offering them not only to Buller but to Buller's chief rival, the famously well-to-do bird collector Walter Rothschild.

The wren was truly something else. It had long legs and hardly any wings. Lyall, perhaps the only man since the ancient Māori ever to see the bird alive, sent word that it ran like a mouse and didn't fly. Taxonomists later examining the museum specimens found its flight equipment all but jettisoned. Its wing bones had shortened, its flight feathers been rendered aerodynamically unfit, its breastbone—to which its major flight muscles would have otherwise attached—withered to nearly nothing. The Stephens Island wren never flew; it ran in fits and starts, under cover of darkness. It had indeed become a feathered mouse.

Which on an island with a cat having nothing better to do made the wren a most appealing sort of game. Not that it was merely one cat doing the killing, as the story usually goes. There was at least a family of them prowling the confines of Stephens Island. (From which they eventually multiplied so profusely that the keepers started shooting them as pests.) Nor was the supposed villain even named Tibbles. (The name seems to have been invented for the sake of good copy.) Nor did he, or she, even belong to Lyall, who was merely the messenger.

Lyle nonetheless did write Travers of the bird's impending doom: "The rock wrens are very hard to get, and in a short time there will be none left." To which Travers responded by raising the asking price in his pitches to Buller and Rothschild. And with the cats doing the killing, and the collectors and profiteers haggling over the carcasses, the little team of conspirators ran through what would turn out to be the first and last of the odd little wrens.

Within a year of discovery, the *Christchurch Press* was reporting that "there is very good reason to believe that the bird is no longer to be found on this island, [and] as it is not known to exist anywhere else, it has apparently become quite extinct. This is probably a record performance in the way of extermination." The *Press* may have been correct, or it may have been another year or three before the last Stephens Island wren stopped scampering about the island. But the tale fell decidedly short on the larger history of what had actually befallen the little bird.

Lost from the tidier bedtime story was the fact that "Stephens Island wren" was itself a misnomer. Fossils of the bird were later to be found all over mainland New Zealand. Stephens Island, it turned out, was not the birthplace of the evolutionary oddity but the final refuge of a once widespread bird driven to the very edge and, through the paradoxical misfortune of its discovery, witnessed at the moment it finally teetered off into eternity.

And so it was, after two hundred million years of splendid, decadent isolation and evolutionary experimentation, capped by six quick centuries of human occupation, that New Zealand limped toward the close of the nineteenth century a hollowed-out shell of life. It had been a siege of unprecedented rapidity and scale, the explosive finale of Oceania's great extinction.

## Chapter 2

## RESOLUTION

IN 1894, THE year in which the cats and collectors of Stephens Island were chasing down the final few survivors of the wren that would be a mouse, the New Zealand parliament was otherwise concluding that perhaps something ought to be done about it. Perhaps it should secure a wild place where the surviving avifauna of New Zealand could huddle in safety out of reach of the plague of foreign predators now bearing down on them. Perhaps not only a place but also someone to usher them there, and to look after them as well.

For this purpose the legislature set aside an island off the country's southwest coast. Resolution Island was 47,500 acres of thickly forested mountains and sheer, forbidding shores, surrounded by a mile-wide moat of cold water. It was the largest of a hilltop archipelago in Dusky Sound, where the glacier-carved valleys of Fiordland dipped into the sea. Resolution was an unpeopled island in a lonely place. No roads led to Dusky Sound. The rare visitor came by ship or, less often, by foot, after weeks of intrepid cross-country tramping through uncharted mountain passes. Through centuries of sporadic intrusions, from Māori moa hunters to prospecting miners, the human footprint on Resolution Island had remained light. More critically, for New Zealand's last flightless birds, it had yet to be infected with the rabbits, stoats, cats, and dogs that stalked the mainland fauna.

Resolution Island, the government decided, would be a good place to harbor the last of its country's most critically walking wounded, chief among them the kiwi, little wingless cousin of the Australian emu; the weka, one of the two big flightless rails of New Zealand yet to be annihilated; and that incomparable waddling parrot of the night, the kakapo. Resolution was to be stocked with the dwindling birds from the mainland, to shelter them against the coming storm of invaders.

In 1894 the government allocated an annual salary of 123 New Zealand pounds, plus a small allowance for building a shelter in the howling wilderness of Dusky Sound, for the man whose task would be nothing less than to carry out the rescue of Fiordland's last flightless birds. For the task it looked to a homeless man.

## The Handyman

Richard Henry was adrift and penniless on the far end of the country, in Auckland, when the telegram reached him, announcing the opening for caretaker of Resolution Island. Forty-eight years old, limbs creaking from a life of hard knocks, Henry better fit the hobo's description than that of the one man in New Zealand to save the kakapo.

Henry's wearied soul was the victim not of rust but of long, hard mileage. He'd been five years old when his fortune-seeking father had shipped the family—seven children and a wife—from their homeland of Ireland to the boomtown of the Australian frontier. By the time the boat had docked four months later, Henry had watched both his mother and his infant brother succumb to sickness, to be buried at sea.

Australia offered the Henry family few sympathies. Richard's father found his hoped-for land of opportunity fierce with competition. John Henry tried storekeeping, engineering, surveying, and architecture on his way to going broke. Two years after landing his tattered family in Australia, he lost his six-year-old son to typhus. He borrowed money to move the family to the blooming port city of Warrnambool; a trip normally figured at a few days took three weeks after a storm blew the ship out to sea.

In Warrnambool, John Henry held together what was left of his family, turning odd jobs as a mechanic and a carpenter, slowly regaining his feet. His son Richard, torn between an inordinate yearning for the outdoors and a sense of his duty to help provide for the struggling family, melded the two impulses into a single impassioned pursuit. At ten years old he was plowing fields and shooting pigeons for food. He took to the woods and fields and swamps of the Australian outback, with a young naturalist's fascination and a gun. Richard Henry apparently spent little time in formal schoolrooms, studying instead Australia's Aborigine hunters. He emulated their techniques, spearing eels with bamboo reeds, climbing trees with bare feet and a tomahawk to reach the possum's lair. The budding bushman paddled the rivers in a bark canoe, fishing and gathering the eggs of wild fowl. By the time he was fifteen, he was bringing home kangaroo meat for the family table.

Henry was twenty-two years old when he took up working in his father's sawmill. Not long after, the saw errantly ejected a block of wood like a cannonball and struck his younger brother Alexander dead. For Henry it was the final blow to what remotely passed as a family life. He fled the painful memory of the mill, disappeared into the Australian outback, and took up wandering as his chief vocation.

Henry would surface now and again, as a jackaroo on a

sheep station, as a hired hand in a backcountry timber mill. But he would invariably get his fill of the foul workmen's crews and steal away to the comfort of the wilds and the companionship of the Aborigine. "They were good-humored, jolly company, full of fun and activity and kindness towards each other," he later wrote. "When I went home to the man's hut at the sawmill and into the surly, overworked, blasphemous company, it contrasted unfavourably with that of the happy darkies."

Henry came to master the art of self-sufficiency by way of many trades, as shepherd, sawyer, carpenter, and boatman, but above all as bushman. Given his druthers, he would be out there among the wild animals, pondering their beauties in one moment if then bagging their bodies in the next.

When Henry pulled up stakes again in 1870, he went this time for more than the average Australian walkabout. He "crossed the ditch," sailing the stormy waters of the Tasman Sea to the shores of New Zealand.

He landed to find New Zealand in the throes of its rabbit plague. The rabbits had by then officially crossed the line from sportsman's bright idea to national scourge, making fools of their liberators and a shambles of the sheep industry. And Henry the ever-handy bushman found quick if unfulfilling work in the Sisyphean task of shooting them.

Whenever Henry bored of rabbiting, he would wander on to a place or trade that better suited him. He did some boat building, some carpentry, more hunting and fishing, all ultimately leading him to open air and the wild creatures of the backcountry. Somewhere along the way, as he let on in his journals, he also met a woman, a woman he hoped to marry. Then, with merely a footnote dedicated to her subsequent rejection of him, Henry hurried on again in his retreat from civilization, in search of the next wildest place on Earth.

By 1880, Henry had retreated as far as men then went in New Zealand, to the southwestern corner of the South Island, on the shores of Lake Te Anau, gateway to Fiordland. To the east spread rolling pastures of sheep and cattle. To the west rose the white glaciers and crowded peaks of the Southern Alps, separating Lake Te Anau from the Fiordland coast, the last frontier in all of New Zealand.

Here he had finally found the place seductive enough to hold him. He had endless work at the sheep stations, where locals came to know him as Henry the rabbiter and Henry the shepherd. In the budding settlements he became Henry the carpenter and Henry the sawmiller. He learned the lake and practiced his sailing craft for pay, as Henry the boatman and Henry the guide.

The seasonal jobs financed Henry's true passion. He would finish his work, pack his sailing dinghy and dog, and shove off on naturalist sabbaticals to the west and wild side of the lake. He would lie by night in camp and listen to "the perfect din" of birds calling from the bush.

Henry's attentions came to narrow on the one bird dominating the symphony. From out of the hills came the haunting, gut-stirring vibrations of a bass drum slowly beating. Henry would set out into the hills, to creep within what seemed a few yards of the twilight percussionist, and release his dog. Then he would follow, scrambling through hill and brush to finally find his dog pinning the giant parrot of the night, the kakapo.

Henry beheld a soft and beautiful bird with feathers the foliage green of the surrounding scrub—an intricate plumage that camouflaged the kakapo as a beaked patch of shrubbery. The bird also came with a pleasantly conspicuous scent. Designed as a note to fellow kakapos, it was now received as a dead giveaway to the land's new hunters. Henry's dog could smell kakapo a quarter mile away; he caught them by the score and relished their tender meat. Henry could only shake his head at the bird's helpless plight. "They are the easiest things in the world to exterminate," he wrote with foreboding. "A few wild dogs would clear the country in a decade."

Henry would cradle the endearing kakapo in his arms and, likely as not, then kill and stuff it. He had become one of the smaller operators in a bustling trade of bird skins, selling off the bizarre New Zealand avifauna to stock the international museums of academia and the display cases of the wealthy.

The killing presented no great moral dilemma for Henry. The virgin hills beyond Te Anau still ran thick with kakapo and kiwi. And ultimately, Henry the collector reported to Henry the naturalist. With every bird he would chase or kill would come hours of admiration and contemplation. Before shooting, he watched and listened. After shooting, he gathered and dissected his victims' gizzards, recording their food habits as a good scientist might.

Such scientists, in Henry's estimation, were too rare. His degree came hard-earned, from days and nights of dirt time in the kakapo's kingdom; he questioned the wisdom of those pontificating from their professor's chair. "When you turn to a big natural history book for information," he wrote, "you find all the fine names and straw-splittings about their classification and species, but hardly a word about their life history or the life of the young ones—whether the mother takes any care of them or feeds them, or if they care of themselves, and what they live upon, though some of those items are of the first importance."

Though ignored by academia, Henry the naturalist became legend among the locals. The wealthy sheep magnate Edward Melland, Henry's part-time employer and steadfast champion, once said of his leading handyman and naturalist savant, "What he did not know about Lake Te Anau & the habits and habitats of its feathered population might truly be said to have not been worth knowing."

In 1883, Henry saw the first signs that his bottomless well of birds had a bottom after all. The ferrets and weasels that had lately been loosed upon the rabbits were rapidly marching westward. "Some one has put ferrets across the Waiau, under Mt. Luxmore," he wrote to those responsible at the Otago Acclimatisation Society. "I was trapping rabbits there and caught two ferrets, so that I think the end of the kakapo has already begun."

For some time thereafter, Henry whistled past the graveyard. He continued his studies and hunts of the beloved birds whose doom he'd predicted. At home in his hut on Lake Te Anau, he would spend his evenings studying heavy works of New Zealand natural history. (That and fishing. He had rigged his line to a bell in the hut that rang when a fish took the bait.) By day he would fill the hut to the rafters with skins and stuffed specimens of the kakapos and kiwis and shorebirds he and his dog had hunted down in the endangered Eden of Te Anau.

To guard his skins while away, he had rigged a diabolical mousetrap, the gist of it recalled with morbid amazement by the wife of his employer, Katie Melland: "Dick took us to see his hut one day, and on entering I was met by an overpowering smell of decayed animal matter and quickly backed out." The smell amounted to a month's worth of dead mice that had been accumulating in Henry's trap while he was away.

"He had a large, square, empty oil tin, with the top cut off, which he had filled three parts full of water," reported Melland. "He made a tiny wooden wheel, like a treadmill, and fixed it across the top of the tin and baited it. The oil tin was sunk beneath the floor of the hut—which was on piles, only a hole cut in the boards to show the wheel. The mouse ran across the floor to the bait, stepped on the small wooden platform, the wheel revolved with the weight of the mouse, round it went depositing the mouse in the water, and was so nicely balanced that it set itself again ready for the next victim."

By 1888, Henry was documenting the demise he had predicted. In the sheep station at Te Anau, where once he could count in one glance sixteen wekas and their broods of chicks patrolling the grounds like barnyard hens, all the wekas were gone. "There was no wanton destruction there, for everyone was friendly to 'the poor weka,'" wrote Henry, "and now that they are gone, everyone without a single exception regrets their disappearance."

Henry watched as the ferrets decimated broods of wild ducklings. And he listened as a countryside once chiming with the calls of wingless birds fell silent. "On the west, from the mouth of the Waiau for 25 miles of beach, there are neither signs nor sounds of kakapo, weka, nor kiwi, where they used to be numerous, but there are plenty of ferret tracks on the beach. Up the creeks in the bush grey teal and blue duck were plentiful, but now they are all gone, and the black teal are rapidly going also, and in all probability will soon be simply a memory of the past."

By the end of the 1880s, within just a few years of the mustelids' arrival in New Zealand, Henry's observations of the birds' demise in the lakes district were being echoed by explorers from across the country's final bastions of wilderness, in the mountains of Fiordland. Weasels and ferrets had been caught and killed within one mile of the sea, far from any point of release. The little predators had crossed the Southern Alps. They had outpaced by many miles of mountain terrain the rabbits they had supposedly been set upon. It raised the obvious question of what, if not rabbits, the carnivores were eating.

"The ferrets and weasels, no doubt, came up the dividing range with the rabbits, but as soon as they discovered our ground birds—our kakapos, kiwis, woodhens, blue ducks, and such like—they followed up the more palatable game," wrote the surveyor George Mueller. "They will continue to thrive until the extermination of our ground birds, now begun, is fully accomplished."

Mueller voiced the bewilderment of explorers across the range, who had returned to their once-fertile mountains to find emptiness, and hunger. "In former times when camping near the head waters of any of the rivers the fighting of the kakapos amongst themselves, and the constant calls of the other birds around the camp often kept people from sleeping. This has all changed now. In the southern parts of the West Coast absolute stillness reigns at night, and there is nothing now to keep the traveller from sleeping except, perhaps (owing to the absence of birds), an empty stomach."

As the birds vanished, Henry began penning articles on behalf of those "perfect fools regarding natural enemies," and with blatant contempt for those now mindlessly obliterating them. "Some of our acclimatisation societies boast of the number of their importations, which may be roughly termed so many nuisances," he wrote in 1889 for the *Otago Witness*, "and now that there is little else to shoot they seriously propose a gun tax, but have not a thought to spare for the preservation of our really valuable natives."

Soon after, in 1891, with the kakapo's demise now imminent, a plan was written to set aside a last resort on Resolution Island. If there were to be a caretaker of Resolution Island, the ultimate candidate would most logically be someone with the skills of an accomplished boatman, capable of crossing the stormy seas of Dusky Sound, someone able to carve a home out of the bush and steeled to the solitude of life in the wilderness. That candidate, most optimally, would also be experienced at capturing kakapos and kiwis. Resolution Island's ideal caretaker described nobody so precisely as it did the eminent naturalist of Lake Te Anau, and nobody wanted the position more than he. No more sawing logs or herding sheep and tourists for a living. Richard Henry would be the bushman who saved the country's inimitable kakapo.

Henry's high hopes soon spiraled. When his unabashed advocate, Edward Melland, pushed for Henry's appointment, the bureaucracy pushed back. There was political infighting among egos; there was talk of abandoning Resolution in favor of Little Barrier Island, farther north. Bureaucrats sniped and both islands sat, while the birds of the mainland continued their tailspin.

Henry had finally glimpsed the life of meaning he'd long searched for, just in time to see it fade from his fingertips. After two years of waiting on the job on Resolution Island, he gave up. He sold his dinghy, *Putangi*, left behind his sanctuary at Te Anau, and headed north.

He stopped occasionally to share his theories of kakapo breeding behavior with the luminaries of academia, to cool receptions. "He thinks more of a classical name than about a curious & wonderful fact," Henry wrote of his meeting in Christchurch with the biologist F. W. Hutton. "He seemed not to take a bit of interest in my story about kakapos but was very anxious to explain to me some straw splitting difference that shifted a bird out of one class into another."

Henry continued north. He tried again at the Auckland Institute, offering his theories of kakapo behavior. Again came the cold hand of the ivory tower, with its polite but patronizing dismissal.

Richard Henry had reached the end of his wanderings. He found himself an aging, unschooled fix-it man with a peculiar passion for the lives of a few odd birds that few others cared to understand. There was nowhere else to go. Biographers John Hill and Susanne Hill would later write of what was to be Henry's final moment.

> Quietly and rationally he carried out his plan. Certain that none would suffer by his action, that he had settled all his debts to the last shilling, and that his body would be unidentified, Henry crept shakily away like a wounded animal to die in a quiet corner apart. He stumbled across a bridge, somewhere, and scattered his last few shillings about, uselessly. Then he took out a six-chambered revolver and shot himself.

Next morning at first daylight, a man admitted himself into the Auckland Hospital. Richard Henry, the ultimate handyman and hunter, had somehow botched the job of killing himself. The first shot had left Henry standing there blinking, the bullet lodged benignly in his skull. He reconnoitered, put the gun to his head, and tried again. The gun misfired. Henry this time took the hint: "The remnants of superstition made me think I had better put it off to see what would turn up."

A week later Henry received a telegraph from Melland, bringing news that he and his mates of the Otago lobby had finally pressured the government into putting a curator on Resolution Island. Two weeks after that, with ship fare wired by Melland and a resuscitated purpose in his heart, Henry was sailing south with hopes of a second life, as curator of Resolution Island, would-be savior of the kakapo.

## To the Rescue

In the New Zealand winter of 1894, the steamship *Hinemoa* delivered Richard Henry to Dusky Sound. But for a transient community or two of gold miners and sawmillers, plus one eccentric old prospector, he was the sole human inhabitant of a water-bound wilderness spanning 150 square miles. He set up shop on a little island to the west of Resolution Island, called Pigeon Island.

Amid the rocky shores of Pigeon Island, Henry found a sandy cove tucked between two sheltering harbors, which would be his port in the stormy seas of Dusky Sound. There he built a house, raised high upon pilings to thwart the periodic stormings of rats. He built a boat slip and a shed and planted a garden in soils mixed with the ancient ashes of the moa-hunting Māori. The forests of Pigeon Island chimed with the birdsong of tuis, kakas, and bellbirds. A cave just beyond the tide line harbored a rookery of crested penguins, where Henry would collect eggs for breakfast.

With his favorite terrier, Foxy, and a young assistant, Andrew Burt, and once again comfortably at the helm of his sixteen-foot dinghy, Henry set out into Dusky Sound, beneath towering snowcapped mountains, through waters breaching with dolphins and whales, headed for the mainland in search of kiwis and kakapos.

He found the wild folds of Fiordland still alive with them. He found the signs of the kakapos' feeding, in the telltale husks and chewings of tussock. In the season of breeding, he felt the hillsides pulsating to the rhythm of their tympanic booming. On their first collecting trip, in May 1895, Henry's little team sailed seventeen miles to the foot of Mount Forster. They returned ten days later (nine of which rained on them) with twenty-six kakapos and a kiwi, and stocked Resolution Island with the first hopes for their future. The rescue was under way.

The rescuers settled into a strenuous routine. They would load *Putangi* to the gunwales with supplies for two-week stints of camping and kakapo catching, and out into the wild sound they would sail.

Henry was careful to study his barometer, to wait out the threatening storm. But once underway, the winds funneling down the fjords of Dusky Sound came quickly and vehemently, forever sending *Putangi* fleeing for shelter in the closest cove. "The steep mountains along the sounds lead the wind, and their many peaks tangle it up so that . . . it is very awkward for a sailing vessel," remarked the understated Henry. "A north-west gale will come down Breaksea Sound, meet the real nor'wester coming in from the sea . . . , and then both go whirling and snorting down together taking a strong current with them."

"Wet and tempestuous" became the standard report in Henry's weather diary. In his first month of residence on Pigeon Island, twenty inches of rain fell; in his first year in Dusky Sound, it rained on two hundred days.

Henry and Burt lived much of their lives in oilskin suits, and wishing they had better. "Our clothes are no use for this climate, and only a load of wet & misery," Henry grumbled, "and the oilskin coat on top of the sweaty wool is a fit finish for a farce in clothing." Seldom the complainer, Henry rued the lead weight of his soggy work boots, "pumping water after half a day in the wet moss & I am certain it would be healthier to go without but for the tender foot of civilization & stupidity."

When Henry wasn't running from a drenching squall, he was shooing marauding rats from his head as he slept and forever swatting the ubiquitous biting sand fly of Fiordland. The maddening swarms of flies had him burning damp moss in his tent to smoke them out. There would come a time when the only salvation for his sanity was revenge. Henry had left his dog tied at camp and returned to find him under siege: "The poor fellow's head was swelled with their bites." He coated his stove's chimney with grease and watched with sadistic glee as the flies glommed on by the tens of thousands. "I was all the evening peeping out through the slit in the door, and greatly enjoyed their difficulties," he recalled. "The woodhens found that sandflies soaked in fat were just to their taste, and they kept up a tapping on the iron that sounded quite musical, because we were sitting in peace for the first time for days."

Back at Pigeon Island, the comforts of home were short-lived. Henry would shave and dry his clothes and begin preparing for the next trip, baking bread and biscuits, preserving penguin eggs, and stocking his food box with stores of bacon and corned beef and potatoes and greens from the garden. Then out into the stormy passes he and Burt would sail again in search of flightless birds.

Upon landing ashore with promising habitat, he would muzzle Foxy, tie a bell to his neck, and, in a routine harking back to his halcyon days at Lake Te Anau, send him coursing and clanging through the bush. And somewhere along the trail, if all went well, at the end of a muzzled nose, would crouch the kakapo.

Henry would lift the bird, as soft as a swaddled infant, and place it in a wooden cage, sending Foxy afield again. Sometimes the team would capture a bird an hour. Other times they would go a day or more empty-handed. But eventually the *Putangi* would sail home to Pigeon Island bearing cages full of kakapos.

The kakapo, Henry quickly learned, was a solitary beast that fought when confined with others. Every bird demanded and thereafter got its own quarters, a fact somewhat comically illustrated by the little *Putangi*, valiantly battling the whitecaps, topheavy with kakapo cages. Home on Pigeon Island, Henry would sometimes feed and fatten his charges in his open-air aviary, before sailing them one last time across the channel to their new home on Resolution Island.

Feeding the temperamental kakapo presented new problems. Ever the individuals, no two kakapos agreed on cuisine. After one of his early captives died in his care, Henry the host pained himself to satisfy the slightest whims of his guests. He hunted and foraged as a kakapo, stooping to harvest the bird's native foods. When the berry crop of the forest petered out, Henry offered bread and potatoes from the Pigeon Island pantry.

Henry's tally of transfers swelled. In July the team spent a week hunting new territory in Cascade Cove (where it rained every day) and came home with two dozen more birds. Three months into his work, seventy-five birds had been whisked out of reach of the predators. By October of that year, the count had surpassed two hundred.

## BOWERS AND BALLROOMS

Henry through his hunting and chasing grew to know the kakapo as no other naturalist of the field or pretender from academia had ever known it. Beyond the more commonly held facts of this singular bird—this solitary, nocturnal, owl-headed parrot that waddled like an elf through dwarf forests of scrub—the kakapo was still queerer by far than any could have imagined. In January 1898, while high on a mountain spine, Henry came upon a network of paths beaten firmly into the spongy earth. They ran for half a mile, interrupted at intervals by round depressions, as if the ground had been stamped by an elephant's foot. Henry measured the depressions at eighteen inches across. He had once imagined these as kakapo dust baths, a hypothesis that now struck him as absurd. No particle of dust stood a chance on a hill that received, by the sodden Henry's estimation, an inch of rain a day.

"So 'dusting-hole' is, I think, therefore, a bad name," Henry wrote. "'Bower' would be more suitable." To Henry these were the ballrooms of the kakapo in courting. It was from these ball-rooms that the booming voice of the kakapo had serenaded him as he lay alone on those late nights at Lake Te Anau. He could now envision the underlying spectacle behind the myste-rious crooning in the darkness. "I think that the males take up their places in these 'bowers,' distend their air-sacks, and start their enchanting love-songs; and that the females, like others of the sex, love the music and parade, and come up to see the show—that is, if they can see the green and yellow in the dark; if not they can tramp along the pathways, listen to the music, and have a gossip with the best performers."

Henry raised questions that hardly occurred to the curators of museum skins but would one day bear heavily on the kakapo's precarious future. Why, he asked, did the kakapo not boom and breed every year? "Can it be that they have curious social laws as mysterious as those of ants or bees—that they have a captain or queen to foresee a season of scarcity or abundance and order their conduct accordingly?"

The kakapo's sporadic breeding schedule, combined with a habit of laying but one or two eggs on average, suggested a spe-cies betting heavily on every chick. Which in turn helped explain their plummetting numbers in a countryside newly swarming with predators.

Henry began to realize the depths of the kakapo's vulnera-bility. Never mind that here was a big, meaty, flightless bird with a fetching scent and eons of ingrained innocence. Com-pared to its eggs and chicks, the adult kakapo was a veritable fortress. The mother kakapo, Henry discovered, as a habit re-ceived no help at the nest. At night, off she would wander, leav-ing behind eggs or helpless chicks. The father kakapo, Henry noted, "won't even keep off the rats while the mother is tramp-ing away for food for her little ones."

Two months would pass before the kakapo chick, so plump and defenseless, was ready to leave the nest. Two months, in a land increasingly prowled by predators, was a harrowing length of time to dodge the inevitable danger.

## THE WEASEL

Reports from the mainland warned Henry that those dangers were fast heading his way. In 1897, a surveying party exploring an overland route to Dusky Sound came back with news that the invasion of the Fiordland coast had begun. "I do not know to what to attribute the scarcity of small birds," reported the expedition's leader, E. H. Wilmot, though he hazarded a guess. "Ferrets or weasels are evidently scattered about, and one of my men says that a ferret paid him a visit in his tent one night."

By then Richard Henry had already begun to suspect that something was amiss in his coastal paradise as well. "I think the ferrets have been down to Supper Cove. In 14 days I saw only one Māori hen and two kakapos . . . We were no distance from the hills and heard no birds at night."

By November 1898, Henry had ferried 572 ground birds, most of them kakapos, to supposed safety. He had overcome the mountains of impenetrable brush, the "roaring fury" of the seas, the rats in his hair, the swarming sand flies, the fickle demands of his captive kakapos. Yet he had underestimated the tenacity of his enemy. In an interview with the *Otago Daily Times*, Henry had once bragged of the inviolate sanctuary of Resolution Island. "When the ferrets come along they will have miles to swim, and they will have, moreover, to battle with fish, gulls, and the tide, and the latter alone is sufficient to disturb the calculations of even good swimmers. On the islands the birds may survive for half a century, and by that time people in every corner of the world will realise their interest and value, and then there will be no fear of their becoming extinct." Such was Henry's confidence when in February 1900 it was summarily crushed with a single blow.

It was then that the fifty-two-ton schooner the *Cavalier*, bearing fifteen tourists, sailed into Dusky Sound. The *Cavalier* met Henry's cutter on the open water and hailed the now-famous naturalist of Resolution Island. Henry shelved his schedule, took a few of the passengers aboard the *Putangi*, and escorted the *Cavalier* on a tour of the sound. He pointed out the mooring place of Captain Cook on Astronomer's Point, pointed them to hiking routes in the mountains.

Before setting sail for home, several of the *Cavalier*'s passengers shared with Henry what to them had seemed a trivial observation. On the morning after mooring at Resolution Island, they had witnessed an interesting little episode, of a weka running along the beach. And bounding fast on its tail was a weasel.

Henry waited for the punch line to what he could only hope was a joke. But the story ended there. The *Cavalier* departed, leaving Henry alone with his living nightmare.

Henry tried to rationalize. There were still many wekas to be found on Resolution Island, a fact that in his experience should preclude the presence of weasels. But there would be no rest until the demon of Resolution Island, specter or reality, was vanquished. The next day, Henry set about making traps. He baited them with fish; he baited them with wekas. The traps lay empty. He mined the bush with the bodies of wekas laced with strychnine. No weasel tracks came near.

Henry held desperately to his hope that the tourists' tale had been hatched as a cruel hoax. "It is a vexatious story & has given me a lot of work," he complained to his supervisor J. P. Maitland. "Why it was started I can't imagine. It spoiled my plans here and upset everything." And after five months of chasing the phantom predator, he was about ready to consider the case closed.

Following a long and dreary July, waiting out an interminable siege of wind and rain, Henry ventured out with the first window of sunshine to check again on Resolution Island. And there he saw, on August 4, in the entrance of Goose Cove lagoon, scrambling upon the rocky shore, the lithe and tubular figure of a little carnivore, hunting in the herky-jerky style of a mustelid. Henry closed to within ten yards of what was now obviously the animal he had most feared, before the weasel caught his scent and disappeared.

For months afterward, Henry struggled with the desperation of a death row inmate. He set traps with dead bait and live, and found weasel tracks as close as ten yards away, but could never touch the creature of his nightmares. There was no escaping the inevitable. Where there was one weasel, there was bound to be more. The channel separating Resolution from mainland had been proven too narrow, the impenetrable fortress was no longer. The predators would keep coming.

Henry started putting birds ashore on other islands; he ventured north through Acheron Passage, to Entry Island of Breaksea Sound, with more birds. But for all the valiant intent, it amounted to the reflex of a man mortally wounded. After a brief vacation in Wellington, Henry returned to Pigeon Island with a broken spirit. Always the meticulous groundskeeper, he could not bring himself to paint his house or tend his garden. His life's calling in Dusky Sound had become nothing more than a job, a chore of tedium and attrition in the face of an unstoppable enemy. "I have not the old interest in it," he wrote in January of 1902, "for I am not expecting a . . . long residence here, on account of that weasel."

A month later, a beaten Henry sent his letter of resignation. "I feel I cannot stay here much longer, so I beg to resign my billet as caretaker of Resolution Island and propose to leave here by the next boat."

Henry would rally once again, but only briefly. He agreed to reconsider, to stay on at Resolution, though the passion would never return. For the remainder of his days in Dusky Sound, he went through the motions. "I am 57 now and have made no worthy provisions so that my pleasant old dreams of getting married will all have to be buried and that will be alright," he wrote. "I have the rheumatics in my hands, often lumbago and this wheezy chest so that I am not half my time fit for work. I have been building a dinghey and I could make myself so tired that I nearly always went to bed before dark."

Out of a rote sense of duty, Henry continued to move kakapos and kiwis, shuffling birds like deck chairs on the *Titanic*. His tally surpassed seven hundred birds. It was a wonder there were any left to move. By then he had concluded that the kakapos and kiwis of Fiordland were on the skids. His favorite old hunting grounds on the mainland had fallen quiet. The refugees stranded on Resolution had become sacrificial lambs to professional killers. "Whatever has been the cause it has been the same everywhere I have been these last two seasons," Henry reported.

Richard Henry would die twenty years later, alone and confused in a nursing home in Avondale. Back in Dusky Sound, his kakapos were abandoned to their fates, the walls of their last little fortress falling before the vandals.

# Chapter 3

## FOX FIRE

NEARLY FORTY YEARS after a beaten Richard Henry
surrendered Resolution Island and his country's tailspin-
ning avifauna to their fate, there began an eerie repetition of history. A lone man in a little wooden boat began crossing treacherous
seas between islands of snowy peaks, on a mission to save a spectacular kingdom of birds.

His name was Bob "Sea Otter" Jones, and in 1947, in a far-
away island wilderness a hemisphere and six thousand miles
north of New Zealand, the sturdy little seaman Jones took up
sailing a twenty-foot dory through the storm-battered archipelago of the Bering Sea, as the resident first manager of the Aleutian Islands National Wildlife Refuge.

Jones had accepted the Herculean task of managing the
Aleutians' sanctuary of seals and seabirds, gathered on a chain
of cold and rocky islands arcing eleven hundred miles, from the
Alaska Peninsula to the farthest American outlier of Attu. Jones's
work environs had a temper to match the squalling tantrums
of Richard Henry's Fiordland. His was the domain of the infamous Aleutian fog that lured lost pilots into mountainsides,

whose peaks occasionally rained boulders of lava and harbored a beastly wind with a name all its own. The williwaw was a meteorological phenomenon born in the icy mountain peaks, a cold, dense slug of air hurtling downward over shore and sea—an avalanche of wind. On the decks of boats bobbing off the coastal swells, sailors would come to fear that certain sudden calm of a hurricane's eye, heralding the rumble of an oncoming freight train. They would batten down the hatches and brace themselves for the williwaw to come roaring, a mast-snapping, boat-flipping force sometimes reaching speeds of 140 miles an hour.

Williwaws and erupting volcanoes, blinding fogs and fifty-foot seas—these were as much a part of the Aleutian experience as the rare windows of sunshine that revealed the most ethereal of landscapes. Those few who could roll with the punching winds and shrug off the tenacious chill found seduction in an Aleutian majesty. One of those was Bob Jones.

Jones had been educated as a biologist at South Dakota State University. The cold, windswept plains of interior North America served as fair training ground for an Aleutian tour of duty. Jones welcomed the stormy moods of the Jekyll and Hyde paradise. There were beaches in the Aleutians harboring bawling herds of northern fur seals by the hundreds of thousands. There were sea cliffs crammed wing to wing with nesting murres and kittiwakes, puffins and auklets, amassing by the millions. Even in the forbidding winter swells of the Bering Sea, flocks of floating seabirds would stretch to the horizons as the bison had once blanketed the Great Plains.

One could only imagine the Aleutians at their wildest, for their heyday had long passed. However stunning the show, the wildlife multitudes that Jones had inherited were in fact the withered vestiges of an epic plunder.

## BERING

In June 1741, the Russian ship *St. Peter*, with a crew of seventy-six commanded by Vitus Bering, sailed from the Siberian shores of Kamchatka in exploration of the North Pacific. Nearly six weeks later, after reaching the Alaskan shores of what is now America, a mysteriously indifferent Commander Bering celebrated the discovery of his lifetime with an inexplicable impatience to head home. After a brief foray along the Alaskan coast, the *St. Peter* weighed anchor and headed back across the sea that would take Bering's name, as well as his life.

Bering's premonitions soon became prophecy. By late September, halfway home across the Aleutian chain, a third of Bering's crew were lying in the hold, joints aching, teeth loosening, faces yellowing with the slow death of scurvy. Bering himself was bedridden with a mysterious malady all his own. With ship and crew at half-mast, the signature williwaw of the Aleutians came crashing. "We could hear the wind rush as if out of a narrow passage," noted the ship's naturalist, Georg Steller in his journal, "with such terrible whistling, raging and blustering that we were in danger of losing masts or rudder or else of seeing the vessel broken by the waves, which pounded as when cannons are fired, so that we were expecting every moment that last stroke and death. Even the old and experienced pilot Hesselberg could not recall among his fifty years at sea having passed through a storm which even resembled it."

On the morning of September 30 a williwaw more ferocious than the last struck the *St. Peter*. "No one could lie down, sit up, or stand," wrote Steller. "Nobody was able to remain at his post: we were drifting under the might of God wither the angry heavens willed to send us. Half of our crew lay sick and weak, the other half were quite crazed and maddened from the terrifying motion of the sea and ship. There was much praying, to be sure, but the curses piled up during ten years in Siberia prevented any response. Beyond the ship we could see not a fathom out into the ocean because we continuously lay buried among the cruel waves. Under such conditions no one any longer possessed either courage or counsel."

The siege extended through October. Under barrage of wind and wave and scurvy, their food running short, minds and bodies unraveled. Corpses began going overboard with nearly daily routine.

Finally, on November 4, dead ahead of the storm-tossed ship arose a mountainous land, mistakenly imagined by the desperate crew as the shores of their homeland. "It is impossible to describe how great and extraordinary was the joy over everybody at this sight," Steller wrote. "The half-dead crawled up to see it, and all thanked God heartily for this great mercy."

Celebration soon turned to panic. Before dawn, heavy surf snapped the *St. Peter's* anchor and began sweeping ship and crew toward the rocks. Sea-hardened sailors ran crying and babbling. Two corpses being held for burial on land but now arousing superstitions "were thrown without ceremony neck and heels into the sea."

In the moment when all braced to be dashed to their deaths upon the rocks, providence intervened. The *St. Peter* rose on the benevolent crest of a rogue wave, lifting the wounded ship and crew over the jaws of the reef and depositing them at sudden peace in a quiet pool before the beach. The wind calmed, a crescent moon shone above a majestic horizon of sandy dunes and snowcapped peaks. The crewmen of the *St. Peter* had just experienced the luckiest moment of their lives. The natives of the land before them, however, had just experienced their most unfortunate.

The next day Steller and a small crew rowed a longboat from the wreck of the *St. Peter* to the beach to begin exploring what the wishful among them were praying was the Kamchatka Peninsula of Siberia. What Steller soon saw told him it was not. The crew was greeted by a company of curious sea otters, mindless of any danger. Steller's suspicions grew with the appearance of a huge dark creature wallowing in the shallows, a mysterious animal the shape "of an overturned boat," its snout occasionally surfacing to draw breath "with a noise like a horse's snort." The sea cow, as Steller came to call it, treated the boatloads of armed men as mere logs of driftwood.

Once ashore the men were besieged by arctic foxes , snapping and barking. The crewmen kicked them, to no avail. They hacked and stabbed the little creatures with axes and knives. The fearless foxes kept coming.

As the survivors of the Bering voyage settled into their new task of surviving on what would one day be named Bering Island, the foxes became the most intimate and incessant reminder that this strange and hostile land was someplace other than home. The foxes feared nothing in the men who now attacked them. Steller and a shipmate killed sixty in a day. And still they swarmed, bold with hunger and mindless of consequence. "When we first arrived they bit off the noses, fingers and toes of the dead while their graves were being dug," wrote Steller. "While skinning animals it often happened we stabbed two or three foxes with our knives because they wanted to tear the meat from our hands." Men learned to sleep with club in hand. Steller continued: "One night a sailor on his knees wanted to urinate out of the door of the hut, a fox snapped at the exposed part and, in spite of his cries, did not soon want to let go. No one could relieve himself without a stick in his hand, and they immediately ate up the excrement as eagerly as pigs."

Mad with vengeance, the castaways resorted to torture, gouging the foxes' eyes, hacking off limbs and tails, and hanging pairs of foxes by their feet to watch them "bite each other to death." For their efforts the men were thereafter haunted at their huts by tail-less foxes and foxes hobbling on three or two legs, advancing with a zombie's resolve.

As the plague of scurvy abated and strengths recovered under Steller's doctorly care, the castaways turned the natives' fearlessness to their favor. Sea otters and fur seals, lounging and napping so trustingly upon the rocks and sands, allowed the men to tiptoe down and club them. The otters became such predictable fare that the men began throwing away the half-palatable meat and collecting hides as poker chips. "The sickness had scarcely subsided when . . . worse epidemic appeared," recorded Steller. "I mean the wretched gambling with cards, when through whole days and nights nothing but card-playing was to be seen in the dwellings, at first for money, now held in low esteem, and when this was gambled away, for the fine sea-otters, which had to offer up their costly skins."

Faced with the relentless slaughter, the otters and innumerable rookeries of seals that Steller had noted at the onset steadily evaporated. Hunters found themselves journeying farther for their meat, trudging miles over rocky tundra to new shores in pursuit of the retreating herds. On the way, they hunted by the hundreds the ptarmigan, the snow grouse of the Arctic, until the ptarmigan too grew hard to find. They chased down the island's strange species of cormorant, a fish-eating bird big enough to feed three and, to its ultimate demise, flightless.

Eventually the men of the *St. Peter* turned their attention to the most tempting mass of meat on the island, which came in the form of a sea beast thirty feet long and some four tons heavy. The sea cow, a gigantic cousin to the manatee, lolled in familial pods along the shallow shores. Steller, ever the naturalist, had been observing them daily from his hut. He would come to know these animals as no other human ever would.

"They come in so close to shore that not only did I on many occasions prod them with a pole or a spear, but sometimes even stroked their back with my hand," Steller wrote. "If badly hurt they did nothing more than move farther away from shore, but after a little while they forgot their injury and came back."

After a string of frustrated bunglings and escapes of wounded sea cows, the hunters eventually honed a crude but lethal technique. It amounted to live butchery, which Steller seamlessly recorded with compassion and chilling candor. "These . . . gluttonous animals keep head under water with but slight concern for their life and security, so that one may pass in the very midst of them in a boat even unarmed and safely single out from the herd the one he wishes to hook . . . Their capture was effected by a large iron hook the point of which somewhat resembled the fluke of an anchor, the other end being fastened by means of an iron ring to a very long and stout rope, held by thirty men on shore. A strong sailor took this hook and with four or five other men stepped into the boat, and one of them taking the rudder, the other three or four rowing, they quietly hurried towards the herd. The harpooner stood in the bow of the boat with the hook in his hand and struck as soon as he was near enough to do so, whereupon the men on shore, grasping the other end of the rope, pulled the desperately resisting animal laboriously towards them. Those in the boat, however, made the animal fast by means of another rope and wore it out with continual blows until tired and completely motionless, it was attacked with bayonets, knives and other weapons and pulled up on land. Immense slices were cut from the still living animal, but all it did was shake its tail furiously and make such resistance with its forelimbs that big strips of the cuticle were torn off. In addition it breathed heavily, as if sighing. From the wounds in its back the blood spurted upward like a fountain."

Finding their stricken mates and family members under assault, the sea cows would surge heroically to their rescue. "To this end some of them tried to upset the boat with their backs, while others pressed down the rope and endeavored to break it, or strove to remove the hook from the wound in the back by blows of their tail, in which they actually succeeded several times. It is a most remarkable proof of their conjugal affection that the male, after having tried with all his might, although in vain, to free the female caught by the hook, and in spite of the beating we gave him, nevertheless followed her to the shore, and that several times, even after she was dead, he shot unexpectedly up to her like a speeding arrow. Early next morning, when we came to cut up the meat and bring it to the dugout, we found the male again standing by the female, and the same I observed once more on the third day when I went there by myself for the sole purpose of examining the intestines."

## THE RUSSIAN INVASION

By January 1742, having recently lost their commander, Bering, to intestinal gangrene, the survivors had come to realize that they would either rescue themselves from their island prison or die a miserable death there. They began to disassemble the wreck of their ship, to build a new boat from its remains. Come August, nine months after their stranding, with their energies freshly bolstered by the meat of the sea cow, the forty-five remaining seamen of the Bering expedition crammed themselves into a forty-foot boat resurrected from the bones of the *St. Peter* and sailed westward for home. Fourteen days and sixty miles later, furiously bailing to keep their jury-rigged lifeboat from sinking, the survivors of the Bering expedition at last reached the familiar shores of Kamchatka.

It had been a titanic feat of oceanic exploration and human endurance. But with the crew's escape came doom for the natives they had left behind. With news from Steller and his shipmates of the Bering Island menagerie—some of them wearing coats of fur worth more than the average Cossack's yearly wage— fleets of Russian fur hunters were soon racing eastward for their fortunes in the Aleutians.

The sea otter—owner of the most luxurious insulation in the animal kingdom, a fur destined to adorn Chinese aristocracy with garments of soft gold—was shot on the water, clubbed on land, and netted in the kelp beds. The second-finest fur in the North Pacific came off of the back of the northern fur seal. Conveniently gathered onshore in great rookeries, the seals were slaughtered en masse. In 1791 the fur hunters killed 127,000 fur seals; over the next thirty years they killed another two and a half million.

As each new shore went empty, the hunters moved on, sweeping east across the Aleutians. Adding muscle to the industrial slaughter, the Russians enslaved the islands' Aleut people, looting villages and maiming resisters. The assault on the sea otters was soon to be joined by an international force of Americans, Brits, Spaniards, and Japanese. They chased the otters across the Aleutians to the mainland of Alaska, and on down the North American coast to the end of their range, in California. By the turn of the twentieth century, upward of nine hundred thousand otters had been mined from the North Pacific; by 1925 an extensive survey of sea otters tallied zero.

Steller's sea cow, its fat rendered for butter, its oil for lamps, its skin used for boats, was likewise assailed to a predictable end. Twenty-seven years after the naturalist's first glimpse, Steller's sea cow was extinct. The spectacled cormorant, helplessly flightless but far less appetizing than the sea cow, lasted almost a century longer, before it too was relegated to a handful of museum skeletons and skins.

When in 1867, Russia sold Alaska and the Aleutians to the United States, the Americans took up whatever slack the Russians had surrendered, clubbing fur seals at the rate of a quarter million a year. An international treaty in 1911 would finally slow the slaughter, while sparing the last few renegade sea otters that had somehow hidden out the siege. But a more lasting assault on the Aleutian wildlife had by then been set in motion.

The fur hunters had decided to hedge their bets. As they emptied the shores of seals and otters, they added to them arctic foxes. *Alopex lagopus*, the bane of the Bering castaways, was the dominant little canid of the far north, making do on the slimmest of pickings in the coldest extremes across the circumpolar world of sea ice and tundra. The foxes' coats, capable of insulating little metabolic packages against temperatures dropping to minus 80 Fahrenheit, transformed from sleek summer brown to a plush winter white or bluish gray. Blue foxes, the trappers sometimes called them. The little fox of the Arctic could be found surviving the winter on deserts of pack ice hundreds of miles from land; single foxes had been tracked covering straight-line distances of a thousand miles; one fox was supposedly seen within two degrees' latitude of the North Pole. Yet the widest wanderer on four legs had not yet conquered the high seas. To most of the Aleutians, which had rarely if ever been bridged by sea ice, the arctic fox remained a stranger. That, the fur hunters would soon correct, shipping pairs of blue foxes to distant islands on the chain.

·The basic business plan had the foxes procreating on their

private islands, the hunters periodically returning with traps, as farmers to harvest the crop. They assumed little cost for tending their foxes, counting on the islands' native birds to serve as feed. For as long as they lasted.

One might have imagined, at first glance, the avian subsidy lasting forever. By the millions and tens of millions, seabirds gathered along the Aleutian arc, drawn by a feast of fish and a once-inviolable refuge. The islands arose on the southern rim of the Bering Sea shelf, where warmer shallow waters met the cold deep waters of the North Pacific trench. The mixing of waters stirred nutrients from above and below to feed a thick broth of plankton, the drifting micro-masses that in turn fed enormous schools of little fish, which fed the innumerable flocks of fishing birds. On the countless cliffs and headlands and boulder fields rising between these fishing grounds, the seabirds amassed. Their numbers swamped however many raptors—in the form of gull, falcon, or eagle—might come hunting. They congregated in protective bubbles barring terrestrial predators behind oceanic moats spreading tens and hundreds of miles wide through chilling seas and mountainous swells.

The inland reaches of these islands served too, as breeding sanctuaries for a special assortment of ground-nesting ducks and geese, ptarmigan, sandpipers, and songbirds. Through the ages of isolation, these islands had evolved unusually large variations of the mainland's song sparrow and winter wren. They had produced a smaller, oddly honking offshoot of the Canada goose, the Aleutian cackling goose.

To these sanctuaries the Russians introduced the arctic fox. The same brazen fox that had sneered at the armed castaways of Bering Island now found itself loosed in a kingdom of sitting ducks. Through the endless days of the sub-Arctic nesting season, mad with birdlife, the foxes gorged. They ate eggs, nestlings, and incubating parents. Into the vulpine maw went the ducks and geese, the ptarmigan, the sandpipers, and the songbirds. Only the sheerest of cliffs and tightest of crevices harbored appreciable numbers of seabirds against the onslaught. During the bleak Aleutian winter, the foxes survived on their summer caches of eggs and carcasses. They combed the beaches and tide pools for odds and ends, hunting crab and urchin and clam, scavenging the occasional windfall of a dead seal or whale washed ashore. Come spring, with the arrival of the nesting multitudes, the bird slaughter would begin again.

By the 1800s, birds that had once blanketed the islands had begun to go missing. The native Aleut people, who had long fashioned the feathers and skins of birds into clothing, in the wake of the foxes found themselves wearing fish skins instead. The midcentury sale of Alaska to the United States brought anything but relief for the birds. Pelt prices soared; fox farms proliferated. In 1913 the United States set aside the Aleutian Islands as a national wildlife refuge, with the curiously conflicted purpose of protecting their world-class rookeries of seabirds while propagating fur-bearing animals, chief of which was the arctic fox. By 1925 the Alaskan islands housed upward of four hundred fox farms, that year shipping thirty-six thousand pelts worth six million dollars. Among Alaska's major industries, only fishing and mining surpassed the fur trade.

The 1930s brought the Great Depression and the end of the Aleutian fur bonanza. Prices plummeted, trappers abandoned the islands. Their foxes, meanwhile, were left to tend the henhouses. The managers of the Aleutian refuge, with their fur factory all but shuttered, their magnificent bird colonies in tatters, were faced with the question of what it was they were now to manage. The first order would be to figure out what they had. Or, more to the point, what they had left.

## MURIE

In the summers of 1936 and 1937, the pedigreed American naturalist Olaus Murie was assigned to take inventory of Alaska's archipelago wilderness. Murie and a team of assistants sailed and surveyed from the Alaska Peninsula to the western Aleutians, dodging hot-tempered volcanoes and trudging through knee-deep snows, on their way to taking stock of the islands' birds and seals.

Murie immediately rediscovered at least one aspect of Aleutian life that had changed little since his predecessor Steller took note two centuries before: The impudent arctic fox as a habit still harbored a baldfaced contempt for anything human. Murie in his Aleutian monograph told of being charged by one of the foxes for the apparent crime of looking its way. "To my amazement it came all the way, ran up to me, poked me in the arm, apparently with bared teeth for it was a sharp sensation, then ran off a little distance."

Nothing seemed beyond the fox's audacity, nor its appetite. In its droppings Murie found crabs, mussels, urchins, moss, beach fleas, crowberries, cranberries, pebbles, birds, other foxes, and human skin (the skin coming from a burial cave of Aleut mummies, some of them torn "limb from limb").

But the fox's hunting prowess was most impressively displayed in its pursuit of the Aleutian birds. "According to the Aleuts, sometimes a fox will catch an emperor goose when it is asleep and has its head tucked under its wing," wrote Murie. "On occasion, too, a fox will stand on a point of rock where ducks are diving and, when a duck is rising in the water nearby, the fox will jump in and seize it while it is still below the surface."

There were few safe harbors in the peripatetic little foxes' empire. "Blue foxes readily swim from one island to another when the distance is not great," he continued. "Sometimes they will attempt this where there are strong tidal currents and are carried off to sea and lost. Foxes also can climb moderate cliffs with ease. Occasionally, one will even leap across a chasm and down to the top of a pinnacle where ducks are nesting, then clamber down the pinnacle, and swim back to shore. Foxes have learned to take every possible advantage over birds, and the birds must nest on sheer cliffs or inaccessible offshore rocks to be entirely safe."

Islands that had once housed great flocks came up empty in Murie's survey. Various species of ptarmigans had disappeared from fox-infested islands across the chain. The Aleutian cackling goose, once abundant across the archipelago, was hardly to be found. Seabirds in particular, so typically crowded in their rookeries, many of them nesting in burrows dug into the turf, suffered spectacular losses. Colonies of thousands vanished. On some islands, Murie found foxes subsisting almost entirely on seabirds, at times heaped in caches tallying more than one hundred bodies. Blizzards of birds—of gulls, terns, storm petrels, and puffins—fizzled to scarce sightings in the foxes' wake.

Murie returned from his Aleutian surveys having seen enough wilderness and wildlife for several lifetimes. Yet his final report rang with warnings of ruin. "Possibly, there are areas where bird colonies are so huge that the Arctic fox has made only an insignificant reduction in the number of birds . . . but, in many other instances, great changes have taken place. On some of the smaller islands the birds have been almost eliminated, and on many islands such birds as eider ducks have ceased to nest, except on a few offshore pinnacles where they can find protection. The cackling goose and lesser Canada goose have become so scarce that it is somewhat doubtful whether they can survive in the Aleutians. If the migration to these islands should cease, these species would disappear from the Aleutian fauna."

Here in the Aleutian Islands National Wildlife Refuge, the wild spectacles on which the reserve had been founded were being extinguished by a foreign fox, with the administrators' apparent blessings. Before anybody would seriously tend to that irony, World War II interrupted. The war would score yet another wing shot on the Aleutians' wounded birds, introducing a predator more potentially devastating than the arctic fox. Yet it would also bring the birds their first champion, and the beginnings of the campaign to rescue them.

## The Kiska Blitz

At four A.M. on June 3, 1942, two Japanese aircraft carriers steamed into position one hundred miles south of Dutch Harbor on the Aleutian island of Unalaska and launched an attack on American forces. A dozen fighter planes and twenty-two bombers shook the town awake, causing minor damages to a radio station and some oil tanks and otherwise triggering the Aleutian campaign of World War II.

Within the next four days, Japanese troops had captured the islands of Attu and Kiska, establishing an embarrassing beachhead on American soil. And over the following fifteen months the two countries would struggle for strategic control of the North Pacific theater, before the Allied forces would finally take back both Attu and Kiska, to close the curtain on the Aleutian campaign.

Between the opening and closing salvos, the Aleutian campaign came to be known as the war fought not against enemy bombers and battleships but against the weather. Pilots flew blind through endless fog into lurking volcanoes. Williwaws thundered from the peaks and slammed planes into the ground. Foot soldiers suffered far fewer casualties from enemy fire than from trench foot and frostbite, insanity and suicides. Those who served for too long developed the glazed gaze of a pithed frog, a signature malady immediately diagnosed as the Aleutian stare and routinely treated with sedation, straitjacket, and a trip stateside.

The Aleutian campaign most infamously came to be known for the battle of Kiska. The Japanese Imperial North Force, which had taken Kiska in the first week of the war, had for months been hunkered down in its tunnels, stubbornly holding out under a daily deluge of bombs and the starving squeeze of an Allied blockade. On August 15, 1943, American and Canadian troops gathered for the massive offensive to finally take Kiska back.

On Kiska's D-day, 34,426 Allied troops, 95 ships, and 168 aircraft descended on the forbidding island, prepared for battle. They stormed the shores to eerie quiet. The first soldiers to inspect the enemy's underground city found it deserted, the only hostility a few parting shots of insult hand-scrawled in bad English on the walls. Trigger-itching Canadians and Americans advancing through the foggy hills mistook each other for the enemy and opened fire. A destroyer hit a mine, killing seventy-one.

The Japanese had orchestrated the greatest of escapes. Two weeks earlier five submarines from the Imperial navy had tiptoed into Kiska Harbor under cover of darkness and, in less than an hour, crammed all 5,183 comrades aboard and slipped homeward undetected.

Dozens had died in the Kiska Blitz, every one of them belonging to the Allied forces, each falling to friendly fire or to the mines and booby traps of an enemy gone home. Kiska's weather itself accounted for the most casualties. More than one hundred soldiers, trudging wet and cold through the tundra, were to be treated for trench foot.

With the embarrassing blitz of the Kiska ghost town, the Aleutian campaign came to its end, the war shifting to new and more celebrated battlegrounds farther south before the atomic obliterations of Hiroshima and Nagasaki in 1945 brought the final surrender. Kiska thereafter became a graveyard of war, pocked with bomb craters, littered with military rubble and wreckage, abandoned again to the birds and the foxes, but also now to a new castaway. The rat, stealing ashore while the warring troops were otherwise busy bombing hell out of Kiska, had quietly established its own beachhead.

## SEA OTTER JONES

Among the troops who had fought in the Aleutian campaign was a young radar officer named Robert Jones Jr. Jones had been one of the first ashore on Adak Island, establishing the base from which the final assault on Kiska would be deployed. What his comrades saw as life on Alcatraz, Jones saw as plum duty in the most beautiful place on Earth. Jones was a rough-and-ready seaman, a salty throwback to the frontier mountain man, invigorated by the cold and lonesome spaces of this Bering Sea wilderness. When his postwar comrades went racing stateside to restore their sanity, Jones found a way to extend his Aleutian tour of duty, as manager of the Aleutian Islands National Wildlife Refuge.

Jones took up where Olaus Murie had left off. His vessel was a twenty-foot wooden dory. It was a sturdy, rough-water craft bred in the stormy New England seas, storied home of the tempestuous nor'easter. The boat was propelled by a thirty-five-horsepower Mercury engine, backed by a spare motor in the hold. Jones filtered gas through a chamois to keep the pervasive Aleutian dampness from the fuel lines. A tarp stretched across the gunwales kept Jones's supplies somewhat dry through the squalls.

Jones in his dory became a master of the Bering Sea, less so by battling it, more so by artfully dodging its blows. The sea dealt harshly with renegades and cowboys, its depths littered with the wrecks of boats far larger than Jones's dory. When the whitecaps were angry, when the scudding clouds suggested storms, Jones took heed. There were days of fifty-knot winds and driving rains, days when the manager of the refuge found himself hunkering in a tent. For the rougher crossings and landings, Jones had himself fitted for a neoprene wet suit, a prototype for the modern survival suit. He would thread his dory through the rocks and breakers, toward the least formidable stretch of shore, and jump to land.

From one island to the next, Jones leapfrogged across his wildlife refuge, taking stock as Murie had, with a growing admiration and gnawing apprehension for the endangered Eden of the Aleutians. He could tell by the look and feel of an island—by the relative stillness where commotion should have reigned—of an island eviscerated by fox. To Jones's mind no island could be whole again until he'd driven the invaders out.

To the embattled islands Jones began bringing assistants and lethal arms. They spread baits laced with strychnine and the virulent Compound 1080. They placed exploding capsules of cyanide smeared with scent lures, serving the curious fox a lethal mouthful of poison. They set leghold traps and fired .223-caliber rifles.

With the first eradication of foxes in 1960, on Amchitka, the largest Aleutian Island west of the Alaska Peninsula, Jones's Aleutian campaign was launched.

For at least one among those in harm's way, it was feared the rescue had come too late. When Jones began his tour of duty, the Aleutian cackling goose hadn't been seen in seven years. He would thereafter confirm the goose missing from every island harboring alien foxes. But what about those few islands without?

On June 25, 1962, Jones beached his dory on the island of Buldir. Buldir was a seven-square-mile cone of rock and tundra in the western Aleutians, farthest adrift of any island in the chain. No would-be fox farmer had mustered the nerve to brave the sixty-mile commute between Buldir and anywhere, or the island's infamously treacherous landing upon arrival. Such inhospitable bearing and an absence of foxes had left Buldir mad with birds. Twenty-one species of seabirds bred there by summer in stunning densities—the most diverse seabird colony in the northern hemisphere. In starkest contrast to the Aleutians' fox islands, Buldir was the picture of ecological purity.

Jones's field notes for his July 25 inventory of Buldir recorded with cool scientific detachment a scene to set the heart skipping. Before dryly mentioning three sea otters with pups, a colony of sea lions that "could well exceed 10,000," a pair of bald eagles, and "very large colonies of pelagic birds, especially tufted puffins and horned puffins, murres, kitty wakes, [and] glaucous-winged gulls," he began by noting a flock of 56 Aleutian cackling geese, of the species otherwise presumed extinct. "They were flying off the high steep sea cliffs. They were apparently evenly distributed around the island."

The haven embodied by Buldir provided Jones many thousands of reasons why his refuge with foreign foxes was no refuge after all. He stepped up his Aleutian campaign. The following year he began poisoning foxes on the islands of Kasatochi and Agattu. And in the spring of 1964, helicopters dropped fifty thousand baits laced with Compound 1080 across the rugged breadth of Kiska Island.

Jones was off and running on his mission to liberate the Aleutian refuge. Foxes started dying, islands birds started living again. But what Jones didn't realize was that the fox was only the first and not the worst of his concerns. To this day nobody knows precisely when the rats finally made their way from the World War II battlefield of Kiska Harbor to the great metropolis of auklets at Sirius Point twelve miles north. Whatever the reason, it was later to be confirmed that even as Jones was ridding what he considered the singlemost threat to the birdlife of Kiska, a greater danger was even then advancing upon one of the largest flocks of seabirds on the planet.

The average rat could go where the cleverest fox could not, to the tightest nooks and crannies of the auklets' underground refuge in the boulders of Sirus Point. What a brown rat could do to a defenseless bird a quarter of its size required little imagination. What a horde of brown rats could do to Kiska's uncountable colony of such birds was anybody's guess. But as of the beginning of 1964, with Jones and the Refuge so keenly focused on fox, it was nobody's immediate concern.

By the end of 1964 there would be reason to reconsider. By then a new and sobering demonstration of the rat's ecological power was under way and on public display six thousand miles due south, back on that archetypal island of endangered innocence, New Zealand.

*Chapter 4*

# CAPE CATASTROPHE

IN MARCH OF 1964, the offices of New Zealand's Wildlife
Service received a report from a little island off the country's
southernmost coast, of an outbreak of rats. The news would
hardly have rated a mention had the island been any other than
Big South Cape Island, perhaps the most precious little refuge
for wildlife in New Zealand.

Big South Cape was a wild enclave of rocky shores and un-
trammeled forests spanning scarcely four square miles. Within
those tiny confines, however, the island still harbored—at the
impossibly late date of 1964—a near-pristine assemblage of the
vanishing New Zealand avifauna. Its forests sang with a signature assortment of fernbirds, bellbirds, and fantails, kakas, parakeets, and kererus. Its headlands were swarmed every summer
with seabirds. And crowning Big South Cape's hallowed gatherings was a cluster of the world's rarest species of animals. .

The Stewart Island snipe was a long-billed member of the
sandpiper family, unusually large by mainland standards. The
Stead's bush wren was a mousy sprite recalling the flightless wren
of Stephens Island, the one so infamously snuffed a century before by Tibbles the cat and company. Big South Cape was the

last refuge of a gaudy songbird named the South Island saddle-back, and the greater short-tailed bat, one of only three mammals ever to have reached the country under their own power.

All of these—plus an assortment of native insects, among them a grasshopper approaching the size of a rat—had once ranged widely over the island nation of New Zealand in its pre-predator history. All had since been driven to the edge of their existence, which was now circumscribed by the few isolated acres of Big South Cape Island.

For all its enduring sense of purity, the island had not gone entirely unexploited. Every spring Big South Cape would swarm with sooty shearwaters, saber-winged seabirds arriving to raise their single chicks in earthen burrows pocking the coastal headlands. And every summer boatloads of Māori would land to harvest them. The Māori muttonbirders, as the hunters had come to be called, would gather the fat, flightless chicks from their burrows by the thousands for their feathers, flesh, and oil. Over the centuries the muttonbirders had managed a precarious balance, a sustained harvest of Big South Cape's shearwater crop, retaining as clear a snapshot of primeval New Zealand as could be found in the twentieth century. But now, so very suddenly, the picture included rats.

Sometime in the early 1960s a ship rat or two, stowed away on a muttonbirder's boat, jumped from ship to shore. The rat then applied its signature skill of rapidly transforming all things edible into rat biomass. In 1964 the muttonbirders returned to find their summer camps ransacked, their foodstuffs pillaged, and their cabins fouled. Big South Cape had irrupted with rats.

The news reached the authorities to mixed reactions. For certain academics and administrators, the prospect of rats invading Big South Cape was a minor annoyance for the back burner. But for those earning their keep in the field, the idea rang of a five-alarm inferno.

## MERTON

Among those hearing sirens was a quiet, twenty-five-year-old wildlife officer named Don Merton. In his young but far-flung career with the Wildlife Service, Merton had witnessed firsthand the aftermath of rats invading ecosystems unschooled in defense. He'd been to both ends of the country and beyond, touring the outer islands, seeing the carnage, hearing the silence. Merton had also been to Big South Cape in its heyday. Three years before the outbreak he'd spent a month there, peering through its magic portal into New Zealand's wilder past. It was the enchanting sort of place whose peril the field man could most appreciate.

Merton had found his life's passion early on, in the birds of New Zealand. From the time he could read, he read about birds. He went to bed with the books in hand. Wherever he wandered, be it forest or farmland, shoreline or swamp, he wandered in search of birds. Growing up, Merton kept a birdcage beside the house and practiced avian husbandry. He brought home orphaned goldfinch chicks and put them under his pet canary to raise. He found a wild raptor's egg and duped one of his laying hens into hatching it. His schoolmates nicknamed him Bird Brain. Merton could hardly argue. He was, by his own admission, "absolutely besotted by birds."

To learn anything about New Zealand's birds was to eventually learn a history of loss. It was to know the sad tale of the moas, those wondrously flightless New Zealand giants that had so recently been reduced to piles of butchered bones. It was also to know the less famous but equally poignant story of the huia, an odd, scythe-beaked songbird with elegant black plumes and endearing displays of affection, whose last official sighting had come just a few short decades before Merton's birth.

Young Don Merton had a hard time fathoming such mindless annihilation. And he would be reminded often that the killing wasn't nearly finished. On the Mertons' little family farm there walked a chickenlike bird with a stiletto beak, a tame but mischievous creature, poking about here and there among the ducks and chickens for an unguarded egg. The bird was a North Island weka, one of the few flightless survivors of the great Pacific massacre. It was a recurring specter of a violent past, and of a mass extermination still under way. Merton said to himself, we must not let them go extinct. More to the point, he thought, *I* must not let them go extinct.

In February 1957, three days after his eighteenth birthday, Merton began his internship as a Wildlife Service trainee. After a stint of surveys, touring the outer islands, he found himself temporarily stationed on the southern mainland, on a wild coastal stretch of the Otago Peninsula, as sole caretaker of the Taiaroa Head royal albatross colony. It was famed as the world's only mainland colony of albatross. And it remained so only through the round-the-clock vigilance of bodyguards like Merton. Here Merton learned one of the essential skills of the New Zealand wildlife officer. He learned to kill.

Ferrets, weasels, cats, and rats came to menace the colony. Merton dutifully trapped them all in turn. While not shooing curious visitors to a respectable distance, he acquainted himself with the leghold trap. He learned how to anticipate his quarry's approach, how to lead it with aromatic baits, how to hide the trap's spring-loaded jaws precisely where the interloper's foot was most likely to step.

But there were many predators, and only one Merton. Shortly after his assignment ended, a feral cat slunk into the unguarded colony. Of the five albatross chicks that had hatched under Merton's watch, three were attacked, two of them killed.

It was a lesson for Merton in the realities of wildlife conservation in New Zealand, a lesson harking back to Richard Henry's crushing defeat at Resolution Island. There was no standing back and letting nature take its course, if the nature of New Zealand was to be saved.

## MARIA ISLAND

Little more than a year later, in 1960, Merton was called to investigate a killing on Maria Island. Maria was a three-acre speck of land fifteen miles off the coast of Auckland in the Hauraki Gulf. A schoolteacher and bird enthusiast named Alistair McDonald had visited the island the year before to find it littered with the freshly mangled bodies of some nine hundred white-faced storm petrels. The elegant little seabirds had apparently been killed by ship rats, which had lately run amok on Maria Island.

The rats were newcomers, thought to have arrived in the 1920s, upon rafts of flotsam and floating garbage from Auckland, or perhaps as late as the 1950s, in boats hauling cargo to build a lighthouse. The petrels, for their part, had colonized under their own power millennia before, instinctively seeking a predator-free place to dig their burrows and raise their young. The eventual meeting between rat and petrel came with predictable results, now evidenced by the many lifeless little forms at McDonald's feet.

By the time Merton arrived to investigate, McDonald had already taken up his own vigilante crusade. He had set his fox terrier loose on the rats, then returned with fellow birders to spread poison. Merton for his part took no chances. Over the next two nesting seasons, he and a volunteer crew of petrel defenders armed with a hundred shillings' worth of the rat poison warfarin, went after whatever rats might remain on Maria Island. Merton then moved on to other patients in New Zealand's emergency room of endangered fauna, unaware until years later the historical import of what he had just done.

Merton, in his trials by fire, found himself resuscitating populations of birds breathing their last. It was timely training for the call that eventually came, of the catastrophe unfolding on Big South Cape.

## Triage

By now it was obvious to Merton what the irrupting rats meant for Big South Cape Island. He and his boss, Brian Bell, had sampled widely across the New Zealand archipelago, camping to the late-night din of birds where the predators hadn't yet reached, and lying in silence where they had. Most fittingly, both had visited Big South Cape in 1961, preinvasion, to find the most prolific and pristine of bird sanctuaries. Both knew what now lay in store with the island under attack. Yet neither man could quite fathom what they were hearing from headquarters.

Back in Wellington the decision makers were demurring, suggesting a wait-and-see approach to the unfolding riot on Big South Cape. Bell's requests for action were denied. Days and weeks went by. Bell and Merton sat captive to the tunes of bureaucratic fiddling, all the while smelling the smoke of their precious sanctuary going up in flames. Five months passed before the authorities relented. Bell and Merton gathered a small team and hurried for Big South Cape, to finally step ashore, as feared, to the aftermath of a biological bomb strike.

They peered into the ransacked houses of the muttonbirders, their mattresses shredded for rat nests. Wallpaper—hung with a backing of flour paste—had been stripped as high as a hungry rat could jump. The wreckage outside was worse. North of the harbor, the epicenter of the irruption, the birds were gone. No more saddlebacks, wrens, fernbirds, robins, or snipes. The ubiquitous bellbirds and parakeets, once so raucous, had been nearly silenced. The trees of the birds' forests had been stripped of bark, bushes had been chewed to the ground. Even the insects, including the flagship giant wetas, had all but vanished.

Bell and Merton quickly sorted through the wounded. The short-tailed bat was given up for dead. All attentions came to center on the three rarest resident birds. For the Stead's bush wren, the Stewart Island snipe, and the South Island saddleback, their last refuge had become their death trap. Their only hope, by Bell and Merton's diagnosis, was to be whisked to a new sanctuary.

The crew started netting what was left of the birds. They herded and grabbed the clueless little wrens and saddlebacks as they hopped about at their feet. They formed lines of human drivers to flush the snipes, netting as they went.

As birds came under their care, the team began scouring the surroundings for rations. They spent nearly every waking hour digging for worms, turning rocks for grasshoppers, and trapping moths by lantern light.

The saddlebacks thrived on the medics' makeshift hash. But the wrens' and the snipes' needs would go unmet. The helping hands were hard-pressed to gather the delicate prey preferred by the tiny wrens, to meet the bottomless appetite of the snipes. Their utmost efforts fell short. The last two Stewart Island snipes died in their rescuers' hands. And the fading wrens were down to a harrowing half dozen.

With the world's last six bush wrens in hand, Bell and Merton made a run for refuge. A navy launch ferried them to the shores of a nearby island. Kaimohu was free of predators and people, with good reason. Its gentlest landing was a gauntlet of twelve-foot swells crashing upon jagged rock. It was the wrens' only hope, if also the rescuers undoing.

As the boat neared the rock, Bell stood poised on the bow ready to leap, with Merton halfway back, relaying instructions to the wheelhouse.

"How close, Don?" yelled the captain.

"Twenty feet," Merton shouted.

"How close, Don?"

"Ten feet."

The boat made a go, the waves rose up in menace. It made another go, and another, like a thread searching the eye of a moving needle. On the final run, as the bow clipped the edge of the rock, Bell hopped off.

A deckhand threw Bell a taut line, then fastened tight the cage of wrens. Hand over hand, Bell hauled the last six hopes for the Stead's bush wren across the chasm and into his arms. Stepping delicately, he carried them to the edge of the scrub and opened the cage. There was nothing more that anybody could do for them now.

It was now Bell's turn to be rescued. He scrambled to the edge of the rock, tied a rope around himself, and into the raging winter water of the Southern Ocean he jumped. The cold shock knocked him breathless. The crew hauling fast on the rope nearly drowned him en route. By the time he was pulled aboard, Bell was red as a lobster and shivering uncontrollably.

Bell would ultimately recover; the wrens he'd tried to rescue would not. Over the next few years crews would return to check up on the little refugees, with dispiriting results. Merton saw his last bush wren in 1967. A single bird was sighted again in 1972, for the final time.

And that was that. In a flash of history, even as the birds' would-be rescuers had looked on, New Zealand and the world had lost three species to eternity, hanging on to a fourth only by a hairbreadth and a Herculean effort. The catastrophe at Big South Cape had vindicated Bell and Merton, the messengers of doom. It had served the ivory tower skeptics a sobering lesson in island ecology, a world where rats would be kings.

For Merton the die had been cast—if not with the dead albatross of Taiaroa Head, if not with the petrel slaughter on Maria Island, then certainly with the storming of Big South Cape's matchless sanctuary. New Zealand's embattled fauna was not to be saved by sifting through the smoldering ruins and praying for survivors. Its rescue demanded intervention against the invaders, head-on and with violence as needed. Somewhere out there more Big South Cape massacres were brewing. Somewhere out there the last of the natives were still clinging to the shreds of their homeland, with alien predators closing fast. And most conspicuous among the missing was the bird that Richard Henry had left for dead.

# Chapter 5

# THE NIGHT PARROT

IN THE YEARS following Henry's heroic defeat, the kakapo had slipped into the murky netherworld of hearsay and phantom noises in the night. The latest wave of killer stoats and cats had chased the last of the kakapos to the highest, coldest, most rugged, and least inviting corners of New Zealand. Hunters and road crews pushing into Fiordland's final frontier would occasionally come back mentioning that they'd heard that haunting heartbeat from the hillsides. It wasn't until 1949 that a team from the New Zealand Wildlife Service, in a Fiordland expedition searching for exotic elk, returned with a report of having actually seen, among other things, two kakapos.

The kakapo by any measure was no longer even remotely that ubiquitous voice of the New Zealand outback, whose drummings and screamings had rudely awakened the explorers in their camps, whose tragically grounded trust had once fed their expeditions with easy meat. Over the next decade, having finally come to realize that nobody on Earth could now reliably locate a single living kakapo, the Wildlife Service mounted a succession of searches. Deep into the mountains and valleys of Fiordland they trekked, scouring those places having recently harbored the birds. During the first nine years of kakapo search, the service found none.

It wasn't until March of 1958, in a high rocky meadow beneath the peak of Mount Tutoko, that a search team of five men and two dogs honed in on the first fresh signs, feathers and chewed leaves and seed-studded droppings. A dog on the trail of what all hoped was the telltale scent of the night parrot came to a pointing halt. The bell stopped ringing. Men came running. And there, in the hollow of a large rock, clueless to the world, lay a sleeping kakapo.

Soon in hand, and understandably grumpy over its rude awakening, this somewhat bedraggled, one-eyed parrot nonetheless embodied history. A live kakapo in hand was an event that had taken half a century beyond Richard Henry to be repeated. The team took photos and released the bird, believing with unfounded optimism that they would soon find more.

Next summer the team returned to Fiordland, its ranks this time including a young trainee named Don Merton. Merton would witness with the others the full gravity of the kakapo's predicament. The first half of the season turned up no sign. From one valley to the next, the team found disappointment. Herds of red deer—sportsmen's imports from Europe—had eaten the kakapo's food. Australian possums and British stoats had eaten the kakapos outright. An unspoken fear began wearing on the crew's resolve. Perhaps they had already seen the last of the kakapo.

Finally, in the wildest remnant of the Fiordland outback, high in a walled-in Shangri-La named Sinbad Gully, the team found hope. They had come upon a network of seemingly fresh foot-worn trails running through the bush. The trails veered this way and that, invariably ending in saucer-shaped depressions, what ornithologists had come to imagine as the dust bowls of the kakapo. Never mind that Richard Henry had long ago and sagely observed that dust was a rarer commodity than kakapos in these rain-soaked hills. Whatever their purpose, the tracks and bowls could only be the work of a singular lineage of bird. The kakapo could once again be imagined.

The following summer another party returned to Sinbad Gully, with cages and baits and resuscitated hopes. They camped high in a mountain cleft bearing encouraging signs and set their traps. On the morning of January 25, they found two of the cages occupied. One contained a possum. The other held a kakapo.

There was no letting this kakapo go. Cradled like a long-lost infant rescued from the wilderness, the precious bird was hurried off the mountain and into a waiting car and driven cross-country. Little more than twelve hours after capture, the world's only known kakapo was ensconced at a farm named Mount Bruce, in what was to become the National Wildlife Center. Fiordland, once a seemingly invulnerable fortress, in the modern era of mainland predators was no longer fit for harboring wild kakapo.

But neither, it turned out, were the makeshift confinements of Mount Bruce. By mid-February, four more birds had been caught and transferred to the aviary, to their ultimate demise. They had suffered their long overland trip. And they would suffer thereafter their captors' ignorance and neglect. The kakapo, solitary by nature, fretted and fought when caged in close quarters. Aviaries went untended, droppings piled high, and sleeping quarters crawled with maggots. Disease soon followed the filth. Before the next Fiordland expedition was through, four of the five captive kakapos at Mount Bruce—four of the only five kapapos known on Earth—were dead.

By 1973, with further searches failing, it had become obvious to a particular few that a single male kakapo decaying in a cage was not the stuff of frontline conservation but a front-row seat to extinction. Don Merton and Brian Bell, looking on from the sidelines, lobbied their superiors for an infusion of fresh blood, namely themselves. That year, with the bulldog Bell eventually extracting the administrative concession, his quiet protégé Merton took over as leader of the kakapo field project.

Both men by then had seen, up close, in the catastrophe at Big South Cape, the results of the wait-and-see approach in an island ecosystem under such assault. They had also seen, in their intensive-care rescue of the South Island saddleback, that patients in such dire straits *could* be brought back from the dead. It would require massive doses of resolve and something else all but missing from the New Zealand countryside. It would require the same missing element whose lack seventy years before had ultimately doomed Richard Henry's rescue of the kakapos: a safe and wild place to keep them.

Merton scanned the country map, looking for sanctuary among the islands. The New Zealand archipelago numbered more than seven hundred offshore islands of varying size, but so few of them remained free of kakapo enemies. Down the line, Merton scratched candidates from the list: too little forest, too many people, too many carnivorous animals from foreign countries. On this island roamed feral cats and rats. On that one lurked stoats and possums.

Merton's finger finally came to rest in Cook Strait, between New Zealand's main islands. In the archipelago of Marlborough Sounds there remained a relatively large unsettled island, one of the few still free of the entire suite of foreign predators. Maud Island, decided Merton, was the temporary sanctuary where the kakapo might begin its long walk back.

If only he could now find one. Merton and his crew made new plans for a Fiordland expedition, ramped up with fresh troops and a new means of transport. To deal with Fiordland's overwhelming immensities, the kakapo rescuers enlisted the helicopter. The days of lugging and tramping over snowy peak and jungled dale were to be replaced by a few queasy minutes of vertiginous flight over the jagged roof of New Zealand.

## SEARCH AND RESCUE

In February 1974, the first helicopter-powered expedition in search of the kakapo set out for Fiordland. It was billed as the last chance to save the species; it was financed more as an administrative afterthought. Merton was given money to search only one location.

Merton chose the Esperance Valley. The Esperence was especially rugged and remote, and it came with a sign, in the form of a single kakapo dropping discovered the year before. Soon after setting up camp, the new expedition found fresh feathers. They left out an apple and came back the next morning to find it chewed by the beak of a bird. They rushed to build a holding pen.

The next night Merton staked out the site of the eaten apple. He sat in the cold, waiting, shivering, until his veins suddenly warmed in a bath of adrenaline. From somewhere close at hand had come a rustling. It came from one side, then the next. Whatever was out there was circling him.

From out of the dark came a scream. It was a grating sound that the birdman Merton could only ascribe to one species. He slipped away and returned with a tape recorder. That evening he captured the voice of the kakapo.

Over the next few nights Merton and the team sat out with his recordings, calling for kakapo. The kakapo responded, and now with an apparently keen sense of mischief. The bird would circle the men in the dark, at times, it seemed, within arm's length. It climbed nearby trees and taunted the men with glimpses of its silhouette. Its curiosity grew. One of the men awoke from his post to the sounds of footsteps scuffling past his head. It became debatable as to who was tracking whom. Every morning the light of day would reveal the trap lying empty.

While the resident kakapo doled out its nightly tauntings, the ill-tempered gods of Fiordland soon came to administer their own. Gale winds and heavy rains and summer squalls of snow pinned the team in their tents, blotting out days of fieldwork.

Merton had come prepared. Even while the Fiordland trolls pummeled his nylon shelter, he continued his kakapo studies from his sleeping bag. He pulled out stacks of library papers he'd packed, acquainting himself with almost everything written on this oddest of birds. Which, as would soon become apparent, meant acquainting himself with that oddest of men who had written most of it. Richard Henry, in his years afield, chasing, catching, and caring for hundreds of kakapos, had provided science with its best portraits of the creature nobody else had bothered to know.

Not that science had much listened. Merton, as a student of wildlife, had learned of Henry as the academics had portrayed him, as an uneducated country bushman playing with rare birds. Now, as commander of the kakapo expedition, lying captive in his tent, Merton was painting his own picture of Henry as something other than the village oddball.

Henry had earned his degrees honestly, in the field. While his contemporaries had taken the far simpler tack of killing the kakapo for profit, he had pioneered the more painstaking art of watching and catching them and keeping them alive. He had first perfected the search, behind the nose of a muzzled dog with sounding bell. He had fed and cared for and safely carried to new quarters hundreds of birds. Henry's translocation protocol—the dog, the cages, and special kakapo diets—would become standard practice for Merton and the new generation of bodyguards picking up where Henry had left off. Ever the meticulous observer, Henry had a rare eye for details large and small. He had rightly warned of the government's release of the meat-eating mustelid clan, as the doom of New Zealand's flightless fauna. He had astutely connected the periodic irruptions of invading rats to the subsequent silence of native birds.

Henry, in his life's defining work, had miscalculated by only one vital detail. He had put all his lovingly gathered eggs in the wrong basket. He had put his faith in Resolution, an island very nearly—but disastrously not quite—out of swimming range of the weasel.

Merton would not repeat that mistake, if ever he got the chance. Henry had captured nearly four hundred kakapos; Merton was still working on his first. The camp's impish kakapo was still at large, and looking to stay that way. For more than a week the bird artfully dodged all pursuers, before finally slipping up. On the morning of March 6, in the cage that had consistently come up empty, the crew now came upon the disgruntled figure of a live kakapo.

It was a disheveled specimen, soaked from the night's rain, a single bedraggled entity representing the future of its kind. Celebrations notwithstanding, it was hard to fend off the sobering doubts over the improbable savior standing before them. Even the irrepressible Merton could muster no more than a thinly disguised elegy. "We may be looking at the last of the species."

Jonathon, as he was christened, at first did little to allay their fears. For a week he refused all offerings of food; his weight began to drop. Merton took to cradling the weakening bird in his arm, hand-feeding him a hodgepodge of camp food until he began to find his appetite. Merton offered Jonathon honey water fortified with vitamins and minerals. He started mixing in the bird's native seeds and berries and foliage. Merton recorded every morsel that went into Jonathon's increasingly busy beak. The menu soon exceeded sixty items.

With Jonathon tenuously thriving, the team went back to the bush for what they hoped would be his mate. It would eventually become clear how remarkably little, in the twentieth century, anybody yet knew about the kakapo, an ignorance succinctly evoked by the daily inspections of empty nets. Among their more humbling blunders, the kakapo chasers had been stringing their nets above the river bottoms running through the valley, in hopes of snagging any kakapos gliding across by night—waiting hopelessly as they would later come to realize for a flightless bird to fly.

That season the team managed to capture (with the help of a tracking dog) one more kakapo, a smallish bird they wishfully named Jill. On April 1 they helicoptered Jill and Jonathon out of Fiordland and over the mountains to the holding pens of Maud Island, to try again what Richard Henry had so heartbreakingly failed to do some sixty years before on Resolution Island. They were to populate Maud Island with as many pairs of kakapos as they could find, then help the kakapos take it from there. They had huge advantages over Henry, of manpower and the wings to fly anywhere the kakapo might still be hiding. They had the disadvantage of not knowing for sure whether there were any more hiding kakapos to be found.

The following November, with the dawning of the New Zealand summer, Merton's team returned for their second major mission to bring the last Fiordland kakapos to safety. They were welcomed in genuine Fiordland fashion, sequentially flogged by storms and swarmed by sand flies. They became more intimately acquainted with the insides of their tents, and the writings of Richard Henry. They played their tapes, heard no replies. And finally, in a high valley of Mount Tutoko, they came to a freshly hollowed-out depression that they hoped was a kakapo bowl. That night they were regaled for nearly six hours by the booming, from not more than twenty paces away, of a kakapo.

After a Christmas break, Merton returned packing a new piece of wartime technology, a night-vision scope designed for detecting enemies in the dark. He sat through the long dark hours, waiting with his scope, reliving Richard Henry's dream, "to come here in the season with the eyes of a cat." And on the third night, at ten forty-seven P.M., with good light and a cooperating kakapo, Merton watched in perfect clarity the phenomenon that had perhaps never been witnessed by human eyes.

What Merton would dryly record in his field book—"Began sequence in upright position, then lowered body with grunts to horizontal"—was in life a spellbinding performance, beginning with a swelling from the kakapo's breast, its head all but disappearing into a balloon of bird. Then followed a bodily quaking and the celebrated song called booming, the sound of a bird sending forth a tireless series of gut-rumbling grunts—its bowl serving as resonating chamber—as one blowing into a bottle.

Merton was watching a kakapo in love song, thumping its tympanic invitations over the yawning chasms and head walls of Fiordland. It was the pulse of the New Zealand outback, and it was still beating, if barely.

With the spell of secrecy broken, the fortunes of the search team suddenly flowered. In Sinbad Gully, Merton and crew heard two more kakapos booming from the ridge above camp, one of which now turned the tables on its pursuers. Wildlife technician Rod Morris was out at midday, retracing his steps from a kakapo bowl, when up the trail came walking a kakapo. Morris stood in shock. The bird kept coming. It strutted to within a couple strides of Morris, stopped and spread its wings, and clacked its bill. It climbed a tree, inspected Morris once over, then climbed down to its bowl and began to boom.

Over the following days, the kakapo observers became the observed. The curious kakapo would walk into their blind to inspect things, to be treated with handouts of cabbage and rudimentary attempts at conversation. It would turn out the lonely parrot was interested in more than small talk. He would perform his dance for the men, swaying from side to side, languorously waving his wings like a butterfly.

He grew bolder, climbing pant legs, perching on shoulders, sometimes capping his ascent with an attempted cranial copulation. The amorous kakapo of Sinbad Gully was baldly hinting that these booming birds the men had been chasing through the hills were the lotharios of kakapo society. And that anything coming close to their bowls was to be considered fair game for either fighting or romancing. And that Richard Henry, once again, had been right after all. Henry in his years of observation and contemplation had astutely interpreted—as Merton, with his high-tech equipment, was now confirming—the kakapo's mysterious midnight antics as that of competitive romance.

The kakapo had evolved a communal mating system that science had come to call a lek. It was a courting strategy more typically ascribed to such birds as the prairie chicken and the bird of paradise. The lek had never even remotely been suspected of a parrot, except, of course, by the one man so many had ignored. The ballrooms that Henry had so fantastically surmised to such rude reception were now the stuff of fact. This oddest of birds—this flightless, owlish night parrot—was proving stranger with every glimpse of its secret life. And the oddball Richard

Henry was to Merton's mind proving himself an ornithological savant.

But the answer regarding the booming bowls now raised another, more disquieting question. Given these randy male kakapos, mounting men's heads and wadded-up sweaters with madcap passion, where were the women? The searchers had yet to find—with the dubious exception of Jill—a female kakapo. It was possible that these odd parrots, from a line of birds known for living fifty years or more, were booming for ghosts. It was possible that these were the last desperate love songs of the living dead.

In February the men returned to the Esperance Valley, where the kakapos Jonathon and Jill had been trapped. They played their tapes and set their traps. And no bird sang.

Their hopes waning with summer's end, they flew on to the razor-edged ridges of Gulliver Valley, making camp high on a precarious ledge. There they finally found their hint of hope, in the freshly dug trails and booming bowls of a courting kakapo. A tracking dog named Mandy immediately caught scent and bounded toward the brink of the escarpment, her trainer, John Cheyne, and Merton scrambling behind. The bell went quiet. Cheyne and Merton came running to find dog and bird at a standoff, the kakapo backed to the edge of the dizzying precipice. Merton lunged and grabbed, gathering tight to his chest what would one day become the most celebrated kakapo of the species.

The bird was not the female so desperately sought, but nonetheless a big, healthy male of a species now numbering a handful. Merton immediately took a special liking to this kakapo, an endearing creature that would lightly grip his fingers as he stroked his head. And soon Richard Henry Kakapo was bound for Maud Island, as one of the last hopes for a species.

Richard Henry would be safe on his island refuge of Maud, at least so long as the stoats and rats were kept at bay. Or so long as his health held out. Nobody knew for sure how old Richard Henry was, but guesses were that he'd been holing up in the hills of Fiordland for half a century or longer. His existence conjured the most tragic of scenarios. Had Richard Henry been singing all these years to an empty theater?

The following summer started poorly for what remained of the kakapos. There were neither sightings nor singing to be heard from the hills of Fiordland. Perhaps the fruits of the rimu trees had failed; perhaps the males had taken the year off to await better days for booming.

The three refugees on Maud Island were faring no better. Jill's condition was deteriorating; Jonathon could not be found. Only Richard Henry appeared to be holding his own, playing games with his captors, appearing hail and fat and all but whistling past the graveyard.

In August the kakapo's sliding hopes on Maud took a lurch for the cliff. In a moment of carelessness a kakapo trainee let his dog off its leash, to see it return gripping a freshly killed kakapo in its mouth. A leg band identified the bird as Jill. A necropsy would later confirm the team's suspicions, that Jill should have been named Jack. But the graver matter now became the question of numbers. The last of the Fiordland line of kakapos now amounted to a single bird in hand. Richard Henry Kakapo, so symbolically named, was now alone.

## Resurrection

Merton, the man in charge of saving the kakapo, was in need of a miracle. From the ominous silence of Fiordland, he looked to one last possible refuge. Stewart Island lay nineteen miles off the southernmost shore of New Zealand's South Island. At 674 square miles, Stewart was New Zealand's forgotten mainland. Populated by all of four hundred people clustered in one little settlement, it was also New Zealand's forgotten wilderness. And to some, understandably so. Tangled with brush, flogged by foul weather, bereft of Fiordland's great snowcapped peaks and post-card prominence, it was certainly not the most glamorous place to look for New Zealand's most enigmatic bird. But Stewart Island did have a history of kakapos, sketchy though it was.

There had been sightings in 1949 and 1951, and one as recently as 1970, the latter of the most bizarre kind. A hunter, sitting in a tree waiting for deer, had looked down to the commotion of a screaming beast running beneath him, a beast he described as a "large dullish-green native parrot." Fast on its tail was a rat. The hunter shot the rat, picked up the parrot, propped it on a branch, and took a picture. He reported the bird to an official with the New Zealand Forest Service, who apparently filed the report and forgot about it.

With such crumbs for clues, and nowhere else left to look, Merton sent his men searching. In January of 1977, half of his team descended on the island's southern shore. Within hours of striking off into the bush, the men found unmistakable signs of kakapos.

Here were the manicured footpaths of a flightless bird, radiating from a network of earthen bowls. Two nights later the searchers heard those bowls resonating the bass rhythms of kakapos in song. These were not the hints of one last kakapo biding its time until the great good-bye. This was a thriving population. On just one hill spread a maze of trails and bowls tended by no less than eighteen vigorously courting kakapos.

Over the next three years, Stewart Island would host both the brightest and the darkest moments of kakapo conservation history. The brightest came in March 1980, when the kakapo-tracking dog Jasper veered onto a scent. Jasper coursed this way and that, then headed out, the clanging from his collar followed by the striking figure of a tall, gaunt man, bearded to the waist. Gary "Arab" Aburn, a seasoned dog handler and itinerant ka-kapo technician, chased Jasper out of sight, to where the clang-ing finally stilled. He returned to camp minutes later, producing from his capture bag a kakapo like none the team had ever seen. It was a slighter bird with finer features, greener plumage, and a sassier attitude. Arab uttered the hope on everybody's mind: "I think we've got a female."

But how to know for sure? None among them had ever seen a living female kakapo. Merton, though, had certainly seen enough of them dead. Between his stints in the bush he had been searching the libraries and museums. He had read every description of female kakapos that Richard Henry had written. He had talked with ornithologists and parrot breeders and had examined kakapo specimens from museums around the world. Poring over all the specimens, Merton had narrowed his focus to one consistent field mark, a subtle distinction in plumage between the sexes. The tips of the male's outermost wing feathers were faintly mottled. Those on the females were not.

Merton, now with mystery bird in hand and television cameras rolling, began fingering through the wing feathers for markings, wishing for all his worth to find none. No markings appeared on the first feather, nor on the second. Down the line he went, one feather to the next, the feathers coming up blank, every feather to the last. "None on that one, none on that one, none on that one, none on that one. Right? Female!" Merton, in a moment he would liken to "touching eternity," cradled to breast the first female kakapo in modern history.

Before the end of the season, the team had found not only

eleven kakapos but also the first two kakapo nests. The kaka-po's imminent extinction was put on hold. Finally there were females in hand, with more birds in the bush—perhaps as many as two hundred on Stewart Island—and at least a chance that the most enigmatic and endangered parrot had not yet slipped into the vortex. But the kakapo, it would turn out, was not a bird to foster prolonged celebrations.

Amid the exaltations of the kakapo's eleventh-hour reprieve would come a small but sobering reminder of the times. Just before the first Stewart Island expedition had decamped in 1977, the team had discovered the droppings of a cat. The presence of cats on the island came as no surprise. They'd had the run of the place since the 1800s (compliments of visiting seal hunters), and their feral population had since flourished with occasional rein-forcements from the town of Oban. But this one, so close to the kakapos' courtyard, raised fears to another level, fears that were soon to be confirmed.

By mid-1980 the crews on their rounds had picked up two hundred more cat droppings. Six of them contained pieces of kakapo. The next season came with cat-eaten carcasses of kaka-pos, many of them wearing leg bands or transmitters fitted by the researchers themselves. There was no telling how many more piles of feather and bone lay undiscovered in the Stewart Island scrub. Less than four years since rising from the grave, the kakapo was now threatening to reverse the miracle.

Merton requested permission to start poisoning the cats, to transfer more females to Maud Island—anything to salvage the unfolding disaster. The months went by, politics trumped en-dangered species, and permission was denied. For Merton it was shades of Big South Cape, again watching the fire from behind the police tape. When at last the authorities relented in May 1981, Merton and his team scrambled to secure the only two kakapo nests known. Around the nests they laid five hundred pieces of fish injected with Compound 1080. They sat in blinds, in round-the-clock surveillance, guarding kakapo chicks through the nights. Through those long vigils crept an air of futility. Either there would forever be a ranger affixed to every kakapo nest, or there would inevitably come an unguarded moment when an opportunistic cat would undo a lifetime of work. Merton saw himself bailing the ocean with a bucket. To his mind, the only option left for saving the Stewart Island kakapos was the same he had faced five years earlier when he had flown Richard Henry out of his Fiordland home.

The decision to evacuate or to stay would reopen the philosophical rift. Those of the hands-off persuasion—indignant at resigning a million-year work of creation to the ham-handed care of humankind—argued for leaving the kakapo at home, to ride out the storm as nature ordained. Merton and those more willing to play God would cradle the kakapo out of harm's way, for however long it should take.

Merton's God squad was granted a token compromise. The team would be allowed to transfer twenty-five birds from Stewart Island, to another recently cleared of cats, an island off the coast of Auckland named Little Barrier. Once again, though, Merton's fears were vindicated with horrific quickness. Before the kakapo catchers could gather their quota, the hands-off policy left the only two nests unguarded long enough for the world's only three kakapo chicks to disappear.

It would not be the last blow to the kakapo's foundering ship. In 1984, with Richard Henry the sole remaining Fiordland kakapo under protection, Merton and his Wildlife Service comrades legally challenged the board of administrators of Fiordland National Park for drastic action. The last few male kakapos that remained in those stoat-haunted mountains, bearing the last of the Fiordland genes, should be captured and sequestered on one of the predator-free islands, if there was to be any chance of saving them. The park administrators, frozen with caution, stubborn with territoriality, ordered the birds left where they were. It was a decision that would dearly cost the kakapo. In the following years the searchers returned to the once-booming chasms to find silence. By 1987 only four males were known to survive over the immensity of Fiordland. Two years later, when Merton and crew returned for one last search, the only signs were a few overgrown booming bowls and the withered husks of leaves chewed long ago. Those who had demanded that the kakapo live out its days in Fiordland had been granted their wish with tragic swiftness.

In a matter of a few short years, the fortunes of the kakapo had gone from dismal to hopeful to desperate. To save the remaining Stewart Island kakapos, now surrounded by a closing ring of predators, a crew of cat killers were sent in with poisons and the duty to defend. They too were defeated. The rest of the Stewart Island clan were eventually either evacuated by the rescuers or eaten by the cats. By 1992 what remained of the species was gathered in makeshift island refuges, all hopes hanging on human-administered life support.

## Chapter 6

## BATTLE FOR BREAKSEA

THE KAKAPO HAD become a fugitive from its own country. It could not be expected to survive for long with so few birds in so few places, regardless of the number of nursemaids tending them day and night. What the bird needed, beyond the intensive care, was a place to live—a safe and fertile place, free of strange predators, big enough to house kakapo in large, wholesome numbers—someplace far bigger than the stopgap sanctuaries of Maud or Little Barrier. As for the kakapo, so it was for so many of New Zealand's endangered oddities. In many cases the habitat was still there, waiting. But the places still free of predators were not. It seemed clear, at least to a small but passionate lot, where the work of conservation needed doing. And fortunately for the kakapo, there was a movement finally rising to the task.

It had begun in November 1976, with a conference convened in the city of Wellington. Over two days some fifty-seven leading wildlife authorities met over the arcane title theme "The Ecology and Control of Rodents in New Zealand Nature Reserves." Ostensibly a discussion on rats, the gathering would serve as a referendum on the state of saving island life from a world of invaders.

The meeting was called to order by Gordon Williams, who as it happened had twenty years before led the first modern expedition to find the last kakapos of Fiordland. Williams assumed the tongue-in-cheek personage of the white-wigged judge, with figurative gavel in hand and murderers on trial. "Arraigned before this very distinguished tribunal today are four species of rodents who are known generally to be rogues and vagabonds and also on occasion to have committed murder," he intoned. "We will hear about their crimes today. To what extent they are guilty of genocide it will be for you to decide at the end."

Williams had rightly anticipated a jury divided over one of nature's most divisive creations. Pesky pest or deadly force? Fightable foe or indomitable enemy? Between those seeking the answers—a collection of graybeard academics and revered authorities, hill-charging recruits and tenderfoot soldiers—lay a philosophical divide.

Williams began by calling to the stand the botanist and foot soldier Ian Atkinson, who in his surveys of offshore islands had found all three of New Zealand's invading rats repeatedly implicated in serious depredations. The first item in Atkinson's deposition was the brown rat, largest of the three and weighing up to a pound. Legendary sewer rat of the cities, the brown rat was primarily a burrowing ground-dweller, invader of New Zealand's streamside forests and—most ominously for those natives clinging to New Zealand's near-shore archipelago—a willing and accomplished swimmer.

Next up was the ship rat, more an agile climbing rodent, sometimes called the roof rat, sometimes found in walls and attics, but at home as well in the wild forest canopy—and especially lethal to tree-dwelling birds.

Smallest of the three was the *kiore*, or Pacific rat, the globe-trotting guest and stowaway of the Polynesian voyagers. Not so eager a swimmer as the brown rat, not so capable a climber as the ship rat, and ultimately a loser in competition with the other two, the *kiore* had been all but eliminated from the mainland of New Zealand, relegated to several dozen outlying islands.

Atkinson argued that all three acted directly as predators of New Zealand's native animals, most conspicuously of its birds. He had gathered a list of formal reports and observations of the carnage. The list ran long and wide. The brown rat was impli-cated in the eating of penguin eggs on the subantarctic island of Campbell, mallard eggs in New Zealand, puffin eggs in Brit-ain, rail eggs in California, and tern eggs in Cape Cod. The big rat's diet also included petrel chicks in Marlborough Sounds, shearwater chicks and parents in British Columbia, and petrels of all ages in the Hauraki Gulf.

The ship rat was known for, or highly suspected of, taking tern eggs on the remote Pacific island of Palmyra, tropic bird eggs on Bermuda, and petrel chicks from Hawaii to the Galápa-gos. It had eaten the thrushes, warblers, white-eyes, fantails, and starlings unique to Lord Howe Island, in the Tasman Sea, birds that were never seen again.

Even the supposedly benign little Pacific rat, the Māori's re-vered *kiore*, under closer scrutiny grew to a giant-killer. It ate the eggs of New Zealand's robins and tuis; it ate the chicks of its petrels and terns. And the most astounding demonstration of the *kiore*'s predatory powers would scarcely have been believed had it not been witnessed.

In the mid-1960s, while studying Laysan albatross on Hawaii's Kure Atoll, an American ornithologist named Cam Kepler had come upon incubating birds dead and dying on the nest. Over two seasons Kepler and crew found fifty such birds. The albatross—five-pound giants with seven-foot wingspans—had suffered gaping wounds on their backs. The sores spread up to seven inches across, holes deep enough for Kepler to peer inside to the birds' ribs and lungs. It wasn't until he visited the colony at night that the cause of the mysterious mutilations came to light. Kepler walked up on a wounded albatross sitting in the open. "I shut out my light and sat down to watch, waiting a few moments before shining the headlight again on the albatross," he wrote in a note to the journal *Auk*. "When I did so, many rats scampered off his back where they had been feeding."

The rats were eating and killing albatross, in that order. Kepler watched the rats—more than twenty of them—swarming upon the bird and feeding from the expanding wound. The besieged bird would turn and toss rats with its bill; others would scramble to take their place. By the next morning, the rats were cleaning the bones of a dead albatross.

So much for the harmless *kiore*. That all three rats of New Zealand were both capable of and practiced at killing birds small and large, Atkinson's review would seem to have left little doubt. He concluded with a fatalistic caution, warning of the inextinguishable threat and irreversible changes to the native fauna, once rats established their beachhead.

From the audience a hand went up. It belonged to Kazimierz Wodzicki, an esteemed professor from the University of Wellington. Wodzicki had some twenty-five years earlier written the book on New Zealand's invasion by mammals, on their decimation of crops and forests and fisheries. But Wodzicki was now oddly assuming the skeptic's role. From Atkinson's global list of rat carnage—a list that would run four pages long—Wodzicki singled out one case that seemed to exonerate the rat as independent agent of destruction. He pointed to Atkinson's example of Kure Atoll's's colony of tropic birds, an elegant, long-tailed seabird whose numbers had apparently waxed and waned more with the weather than with the rats.

"The essential point is that the population did not really diminish," said Wodzicki. "There were years when the population went down but with favourable weather conditions subsequently the tropic birds were not molested. Generally, over a period of 10 years, the kiore had an impact just like that of wolves and hares in Europe."

With his wolves and hares, Wodzicki had invoked a textbook case of the nascent field of ecology, of predator and prey populations oscillating gracefully through cycles of feast and famine, the two undulating in choreographed counterbalance to neither's ultimate harm. According to Wodzicki's continental analogy, the rat-borne fears were overblown.

But there was something disturbingly askew in the professor's logic. These were not wolves and hares, perfecting their dance of death over the eons, but rats and birds thrown together as strangers in a last-minute hiccup of history. These were contests being waged not in the sweeping geographies of the continents but in the confined arenas of little islands. Wodzicki's theoretical swipe at the worriers had revealed a fundamental divide of ecological worldviews, dividing those who saw such invasions as a scientific curiosity deserving further study from those who saw them as a wildfire to be fought for dear life.

That little divide would rapidly widen. Next up was Brian Bell, there to retell the story of Big South Cape. If Atkinson had left any doubts about the end point of rats loosed upon islands of innocent birds, Bell was determined to let the evidence of Big South Cape forever vanquish them.

Bell briefly reviewed the pre-rat history of the Big South

Cape tragedy, describing the seasonal tradition of the mutton-birders and their innocuous little settlement of cabins, the relative lack of human inroads by means of fire and clearing—in essence, the enduring balance of relative purity of an island standing whole amid a contrasting century of mainland New Zealand's biotic impoverishment.

"It was the birdlife," Bell said, "its variety and abundance, which first took the visitor's attention." When he and Don Merton had first stepped ashore, the forest of Big South Cape had clamored so conspicuously with the song of New Zealand's emblematic natives—its parakeets, bellbirds, fernbirds, and robins. Here also one could still find the rarest of the remainders, the South Island saddleback, the Stead's bush wren, the Stewart Island snipe, and the greater short-tailed bat, the last survivors from the bygone era of New Zealand's innocence.

Then came the crash of 1964—the reports from the mutton-birders, arriving to find their camps ransacked by rats; Bell and Merton's belated reconnaissance to find the forests in tatters and the birds obliterated; and the tragically delayed rescue, wherein Bell and Merton watched as two species of bird and the incredible crawling bat were lost forever.

Muttonbirders and bird-watchers would eventually come together, with poison to knock the rats back. But by then the damage had been done. Said Bell, "Since the irruption has passed and the rat population has reached a more stable level, bird populations have settled to a level one would expect in mainland situations where rats have been present for a long time." Big South Cape was no longer a wild relict of primeval New Zealand, but a mere miniature of the eviscerated mainland.

In closing, Bell could not resist a parting shot at those who had fiddled while Big South Cape had burned. He had not forgotten those five agonizing months that he and Merton had been kept waiting for permission to mount a rescue. "The Big South Cape irruption is now history, but what has been learnt from it?" he asked. "Unfortunately less than one would like because, despite the request made by the Wildlife Service, no research worker could be found at the time to study the rat irruption and its effects." And little was being done to prevent the repeating of that history. Fishing boats were still anchoring close to rare sanctuaries; rats were still jumping ships and crawling down anchor chains. There were more Big South Capes waiting to happen.

And for some, that was apparently OK. In the Q&A that followed the day's briefings, Bell's rendering of the Big South Cape tragedy came under fire. "In my view there were actually two forces acting from the beginning," said Wodzicki, taking the first jab. "The habitat of all these islands had been seriously deteriorating through both fires and the construction of buildings and tracks. My impression is that the birds were previously in an endangered state in which only one more stress was sufficient to tip the scales."

"I think you may have misunderstood me," the junior Bell jabbed back. "I was trying to emphasize that there had not been very much modification. The birding on South Cape is very limited in area and the modification caused by the erection of homes is very minor."

Wodzicki: "My point is that, although the rats were a very important new factor in the Cape, in a healthy population at least some of the other species would become adjusted to them."

"Some of the species obviously did," answered Bell. "But unfortunately some did not."

Unfazed by the gravity of Bell's counter, Sir Robert Falla, considered perhaps New Zealand's senior ornithological authority, joined the attack. Falla had not been consulted before the Big South Cape rescue parties had finally gone in. He had objected to the use of poison afterward. Insiders knew Falla as no fan of the Wildlife Service, for which Bell now worked.

"The hypothesis I am putting forward is that this could have been a classic irruption of a population already present for a long period but at a low level on at least one of the South Cape Islands," said Falla. "There is a difference between this theory and the theory of a fresh invasion of healthy rats running ashore along the mooring ropes of muttonbirders' boats. On the acceptance of that second theory of the rat situation the management reaction is to say, 'My God, we have got to do something'. If you accept the former hypothesis you are justified in saying we do not need to do anything."

Bell could only shake his head at the lunacy of the logic. He and Merton had already seen firsthand what the "we do not need to do anything" hypothesis had done for rescuing Big South Cape, and it could be summed up by three irreplaceable living species eternally rendered to museum cabinets.

The conference reports to follow offered little in support of a truce with the rat. More details mounted on the woefully inadequate defenses of seabirds in a new world of rats, of mass killings and lifeless bodies with craniums excavated, of entire colonies cleaned to the very last chick. There followed news on the disappearance of New Zealand's giant weta—an insect larger than a sparrow—which had been swept wholesale from northern forests in the wake of the brown rat. Another New Zealand specialty, the reptilian oddity known as the tuatara, reported the herpetologist Tony Whitaker, had been all but wiped from the mainland and chased by the *kiore* to the country's last uninvaded islands.

"One thing that disturbs me," interjected Wodzicki, searching again for other culprits, ". . . why did [the rats] wait until our time to attack the tuataras . . . ? If kiore were present since the [Europeans] arrived, how could the tuataras have survived?"

Whitaker's retort painted a frightening scenario. "The tuatara is a very long-lived animal," he said. "Estimates of its lifespan range up to two hundred years. I do not think you could suggest that the introduction of rats would stop breeding of tuatara overnight." Whitaker was implying that the last of the world's tuatara amounted to ancient, aging relics, individually immune to attack owing to their overwhelming size, yet collectively doomed to extinction by rats that were devouring every last one of their young. Said Whitaker, "The fact that we are finding no animals less than two hundred millimeters in length could well explain this."

The second day of the conference moved to the question of what, if anything, could be done about the rats. Hopes were scarce. There was talk of spreading chemicals that would render the rats sterile—an approach entailing months of vigilance and tiny odds of success. There was a proposal to fight the island rats with stoats—those same little foreign weasels that had helped drive New Zealand's native avifauna to the islands in the first place.

The war on rats offered at best an eternal struggle, at worst a lost cause. In his closing of the conference, chairman John Yaldwyn summed up the general mood with what would become an infamous statement of surrender. "Nothing that has been said this afternoon, even the use of stoats for control, would make me think differently," he said. "We have control methods, and methods for reducing populations, but complete extermination on islands is remote or at least a very very difficult thing indeed."

## RAT FIGHT

For most in the audience Yaldwyn's waving of the white flag seemed a sensible concession. But for at least one, it bordered on insult. The junior wildlife technician Bruce Thomas had sat through the meetings as more of a wide-eyed spectator, listening as the big guns boomed. But now, to hear it all end with this pessimistic talk of remote possibilities struck an indignant nerve. Hadn't they all just heard, as he had, about Maria Island?

Earlier in the meeting the noted ornithologist Sir Charles Fleming had raised the question of Maria. He had been intrigued by news of rats having somehow been extinguished from the little island years before. Said Fleming, "This is the first example I have heard of the extermination of a rat population by any control measure."

It was Don Merton who sixteen years earlier, after investigating the slaughter of some nine hundred petrels on Maria Island, had helped the bird-watcher Alistair McDonald spread rat poison there. Surveyors would later return to discover that the island's rats had been more than knocked back. A handful of hardy volunteers, with a seat-of-the-pants strategy and a pauper's budget, had wiped every last rat off Maria.

It was only one three-acre island in an archipelago of hundreds far larger, yet the finality of the outcome had gotten the upstart Thomas thinking heretical thoughts about the otherwise invincible rat. Thomas had by then familiarized himself with the enemy more intimately than most. Raised on a dairy farm in the New Zealand countryside, he had grown up in the company of rats—rats that came scrambling for the leavings at the pigsty, rats that swarmed upon the stream banks where young Thomas wandered. Thomas was at turns fascinated by these wild invaders and keen on seeing them dead. He would sit quietly and watch the rats going to and from their holes, darting here and there to steal a mouthful. He would bait the doorways of their burrows with milk curds, shoulder his slug gun, and, with his fox terrier at his side, wait for the rats' inevitable kamikaze dash. Sometimes the dog almost beat the bullet to the target. But invariably there was always one more rat where the last one had come from.

After high school Thomas came under the wing of a biologist named Rowley Taylor. Taylor, as it turned out, had an unusual respect for rats too. His interests had gravitated early toward an animal that others in the profession had passed up for more charismatic subjects. Island to island, across the New Zealand archipelago, Taylor had watched and trapped, skinned and identified rats. It amazed him how few biologists could, or even bothered to, distinguish one species from another. By the time he was thirty, nobody knew New Zealand's rats better than Rowley Taylor. He admired the rats' talents—their toughness and intelligence, their wariness and agility and explosive fecundity, their willingness and capacity to eat anything that couldn't eat them first. But at the time of the Wellington rat conference, knowing what he knew of those talents, Taylor could not argue against Yaldwyn's bleak prognosis for defeating New Zealand's most indomitable pest.

Taylor's protégé Thomas, on the other hand, was yet too young to be intimidated by such odds. While his elders were nodding their heads in unison, conceding defeat, he had already begun imagining a preposterously more hopeful scenario. Thomas had ideas for ridding an island of rats that would dwarf by leaps of magnitude the serendipitous little campaign at Maria Island.

Two years before the Wellington meeting Thomas had

accompanied a biological expedition off the coast of Fiordland. Breaksea Island was 440 acres of steep and forested rock separated by more than a mile of rough water from the nearest harbor. In all its wildness and isolation the island was being considered as a potential sanctuary for the ailing kakapo, before the surveyors reported back. Thomas and his companions were to find the would-be fortress of Breaksea overrun by brown rats. The rats would summarily scuttle any plans for bringing kakapo to the island. But Thomas, watching the impertinent hordes scurrying through the scrub like rush hour pedestrians, began entertaining a more heretical idea. Why not just rid Breaksea of every last rat?

Unknown to Thomas at the time, there was a discovery under way in an overseas lab that was about to elevate his fantasy to the realm of possibility. Agricultural chemists in the United Kingdom had come up with a new poison.

## The Sleep of Death

Since the 1940s there had been a global campaign against the rat as crop pest and urban scourge, and it had largely been fought with massive doses of a chemical named warfarin. Warfarin was a poison of rather benign origins. It was an anticoagulant, a thinner of blood. It had originally served as a human treatment for thrombosis, an overclotting of the blood, a precursor to stroke and heart attack. When lab workers had subjected lab rats to high doses, though, they had died. The minor and sundry repairs of little broken vessels that constituted the daily workings of healthy bloodstreams became, with an overdose of warfarin, the unstoppable leakage of lifeblood. Victims of too much warfarin died the death of a thousand cuts.

It did not take great leaps of imagination to see the darker utility of this lifesaving drug. As a killer of rats, the anticoagulant approach offered immediate advantages over the leading chemical weapons of the day. Strychnine, arsenic, thallium sulphate, and zinc phosphide, among others—acute poisons targeting brain and nerve—produced fast and sometimes violent reactions in their victims. Whatever few rats somehow survived—and there would always be those few—learned a lesson never to be forgotten, or repeated. Those that watched the tortured writhing of their comrades learned as well. A rat with the look and smell of danger burned into its memory was thereafter an invincible rat. It would sidestep and dodge, hunker and wait, until the poison-bearing enemy grew weary and decamped. Then it would gather its colony mates, and in true rat fashion they would restock their pack with a wiser, warier, tougher force of rats. Warfarin, to the contrary, gave few such warnings.

Warfarin acted slowly and quietly, producing no stricken seizures or terrorizing spectacles. The poisoned rat fell weak and lethargic, too far removed from its lethal bite of food to connect the dots of cause and effect. It commonly succumbed, by outward appearance, as one dying in its sleep. Neither victim nor onlookers ever knew what had hit it. And if by chance the untended dog or child happened upon stray bait, there was a ready antidote, as simple as a prescribed dose of vitamin K.

By the 1950s warfarin had become the world's dominant, if something less than perfect, weapon against the rat. Death by warfarin required large doses, delivered over several feedings. A big eradication campaign required heavy labor and lots of expensive man-hours. More crucial still, if the bait ran out before the rat's luck did, the rat survived.

And that which didn't kill the rat made it stronger. Over time those rats surviving a warfarin attack grew immune. Within a decade of warfarin's international deployment, rats in the United

Kingdom, Denmark, the Netherlands, Germany, France, and the United States were shrugging it off. Brown rats, black rats, and house mice were proving impervious. Survivors were rapidly passing resistance to their pups. With the mutation of what amounted to a single molecule, the rats had begun turning the warfarin war in their favor. The eradicators were now facing a monster of their own design, a global phenomenon that had come to be called the superrat.

When the international alarm was raised regarding the rise of the superrat, the young American Dale Kaukeinen was among those sent to scout it out. For his research project as a graduate student of biology, Kaukeinen traversed the country in search of superrats. In twenty out of fifty cities he visited, he found warfarin resistance. Soon thereafter the U.S. government declared quits on its backfiring warfarin campaign.

Meanwhile the chemists in their labs had been scrambling to defeat the mutant molecule, searching for the superrat's kryptonite. They tinkered with their formulas, cherry-picking the most virulent strains of warfarin's anticoagulant compounds. Finally they hit upon a new mix of molecules that could invariably swamp the most resistant animal's clotting mechanism. One of those compounds they named brodifacoum. Brodifacoum's toxicity surpassed that of warfarin by a hundred times. A single gram would kill the biggest brown rat. With brodifacoum there would be no need for multiple doses. There would be no escapees, no wounded survivors to spread their resistance and revive the masses. It was now a matter of getting the rat to eat it.

Kaukeinen, fresh out of grad school, was enlisted in the international team assigned the job of readying brodifacoum for the field. At a lab in North Carolina, his first task was to find the bait package that would stand up to the weather and irresistibly lure every rat. He would put out placebos, packaged in various forms and recipes, and watch through one-way mirrors and hidden television cameras. From his blind he would come to acquaint himself with an animal supremely schooled for the scientist's little games of cat and mouse.

He began by mixing his baits with food preservatives, for better shelf life. He added innocuous preservatives, of the kind found in every commercial loaf of bread. His rats would have none of it. He added a tiny amount of insecticide, to keep the ants and cockroaches away from his baits in the field. The rats wouldn't eat it. They were detecting Kaukeinen's additives in parts per billion—as one might taste a drop of chlorine mixed in a fourteen-thousand-gallon pool. Having been chased through the eons as prey for the predatory masses, the rat had developed wariness to degrees hardly fathomable to the human senses.

But upon finding a food they approved of, Kaukeinen's rats dove in with abandon. They would eat their fill and stash the rest, hoarding food far in excess of their appetites. Kaukeinen would dig up their burrows and find single larders stocked with twenty pounds of rat chow. It struck Kaukeinen as neither gluttony nor greed, but a shrewd survival strategy. "It's something ingrained in them," said Kaukeinen. "That gene to hoard. I can imagine when that probably came in pretty handy. You have a natural calamity, or human disturbance, your regular food source gets cut off. You didn't have that nice odor trail to the dumpster anymore. But you had this big slug of food back in the home nest. So not to worry. They didn't even have to go outside and expose themselves to danger to keep eating."

Then there was that infamous rat talent for evasion. Kaukeinen would occasionally notice that one of his subjects was missing from the lab (or from his shed at home, where, truth be told, he sometimes kept the enemy as pets). The escapee would disappear, and Kaukeinen would assume he'd seen the last of it.

He would dutifully put out baits and traps, to no avail, and return to his business. But eventually Kaukeinen would come upon little signs—a track here, a dropping there—of a midnight visitor now watching *him* by day.

Among the most admired of Kaukeinen's subjects was the Kleenex Thief. The lab workers had begun to notice a box of tissues mysteriously emptying sometime between their departure at night and their morning return. One night Kaukeinen stayed at work and waited, looking through a one-way mirror into his lab, illuminated with a red light invisible to rodent eyes. At the appointed hour, the Kleenex Thief made his appearance. Scampering across the suspended ceiling, shimmying down an electrical conduit, scurrying across the tile floor, jumping up to a chair, and beelining across a cabinet—straight to the Kleenex box the rat proceeded, yanking out one single-ply tissue, then retracing his steps through the obstacle course, prize in mouth. This process the rat would continue to repeat in a tireless rendition of capture the flag. The Kleenex Thief, it now became clear, was no he but a maternal she, dutifully building her nest.

Even the one supposed rat weakness, its legendary nearsightedness, Kaukeinen would find to be badly and perhaps wishfully overestimated. He kept an outdoor colony of rats, on which he regularly spied. He would watch the rats as they were watching their world. He watched as visitors approached the rats' pen. The visitors would invariably leave believing the pen empty. Strangers would get no closer than a hundred feet before the rats would scatter. To see if the rats could be surprised, Kaukeinen had his lab technicians sneak up guerrilla-style. The rats busted them every time. Beady eyes or no, the rat had an uncanny knack for seeing its enemies coming, with good evolutionary reason. These were eyes sharpened to an acute edge by the eternal threat of death, honed to exacting detectors of motion by millions of years of foxes leaping from the grass and hawks stooping from aloft.

Inevitably the games of hide-and-seek and play fighting would end. Kaukeinen would be forced by protocol to proceed to the final step of the experiment, adding to the innocuous baits the lethal dose of brodifacoum. He hated the death. He had come to see in his adversary a soul of admirable wits and resolve. To quiet his conscience, he reminded himself that this was a creature whose kin, the black rat, had once carried the fleas that had carried the bacterium of bubonic plague, which four centuries earlier had swept a third of humanity to horrific death. He kept in mind, as he killed, that this was the rat whose kind was even now eating a fifth of the world's crops, adding filth to his country's foodstuffs, biting and sometimes maiming thousands of his cities' poverty stricken. He had seen, in his own country, homeless people and kids in cribs with toes bitten off. He was reminded daily of the rat's ferocious potential by his own forefinger, left forever numb from one bone-penetrating strike of an incisor.

Brodifacoum, registered in the United Kingdom in 1978 under the trade name Talon, would quickly replace warfarin as the leading weapon in the anti-rat arsenal. Back in the world of island conservation, the poison brought new hopes. Brodifacoum promised quicker death. It meant dead rats with fewer bags of heavy bait hefted into the wilderness. It raised the unthinkable possibility of eradication. Within a year of its arrival on the market, biologists were deploying brodifacoum on tiny islands off New Zealand's coast, testing the new poison in tandem with their more familiar snap traps and other poisons.

Among those most keenly following the developments was Rowley Taylor. He had once been among those accepting the party line, of the futility of fighting this formidable little animal on anything more than the tiniest offshore islands. But with brodifacoum now in the battery, the little victories beginning to amass, and a developing appreciation for the workings of rat society, the odds, to Taylor's mind, had fundamentally shifted.

The rat profile that Taylor had gleaned from labs like Kaukeinen's—and the endless flood of medical and behavioral research on the brown rat's domesticated model, the white lab rat—suggested a creature of complex social skills. The apparent chaos of wild-rat society, as inferred from the occasional glimpse of naked tails scattering before the flashlight beam, was, under closer scrutiny, a well-structured society of leaders and followers, rules and protocol.

Rat society was dominated by big males. The basic fact of their survival to adulthood, in the notoriously fast and ephemeral life of the rat, implied a degree of savvy, a calculated balance of daring and caution, a model to be emulated. The successful rat had evolved as nature's ultimate neophobe, a creature supremely and justifiably suspicious of novelty. Exploring new objects willy-nilly and mindlessly gobbling new foods were behaviors that characterized those seldom living to bear offspring. Mechanical traps, manufactured baits, poison pellets—all underwent careful inspection, most commonly led by the dominant males.

Once the new food or shelter had passed the sniff test of the colony's reigning kingpins—once they'd sated themselves and piled their larders to their greediest heights—the word went out. The message went forth on the scent of rat breath, passed like gossip in the myriad nose-to-nose meetings of the colony's cohorts. The message was "There's food back there, it's good, and the king's had his fill. Let's go."

Following such protocol of rat society, Taylor sketched out his strategy. He would forgo the standard broadside in favor of the surgical attack. He would use just poison and the rats' own intelligence network to defeat them.

## "You're Crazy, Thomas!"

Taylor now needed a target; his protégé Bruce Thomas knew just the place. Thomas had been itching for a shot at clearing Breaksea Island since his rude welcome there by rats a decade before. But both knew better than to aim their first stone at Goliath.

Thomas organized another research trip to Fiordland's Breaksea Sound, this time with Taylor to have a look at one of the candidates, a twenty-two-acre dome of forest called Hawea Island. It so happened that on that boat sailed their boss Richard Sadlier, director of ecology for the Department of Science and Industrial Research. It soon became clear that the would-be rat busters had more than rodents in the opposing camp.

As the three floated past Hawea, Thomas pointed and said to Sadlier, "There's rats on there, Richard, and that's where we should start."

Sadlier scoffed. "Getting rats off a forested island like that? It's impossible. You couldn't do it. You couldn't get rats off a forested island like that. It's impossible."

"Well, you know," Thomas answered, "I think we could, Richard. But what do you think, Rowley?"

The laconic Taylor stood leaning on the rail, staring at the little island, then quietly declared, "Yeah, I think we could do it."

Thomas then turned around, facing the looming profile of Breaksea Island, its monstrous heights wrapped in a band of mist, and dropped his bomb. "But *that's* the island I really want to do."

Sadlier looked at Breaksea and all but jumped out of the

boat. "Rats off that! Rats off that! You're crazy, Thomas! You're absolutely fucking crazy!"

Thomas again turned to his mentor. "What do you think, Rowley?"

Taylor panned the length of the island's forbidding ruggedness, contemplated for a bit, and nodded his head. "Well, it'll take a lot more time and money to set it up, but yes, I think we could do it."

In March 1986, Taylor and Thomas and a handful of volunteers landed on Hawea, for what was to be a live rehearsal for their big show on Breaksea. They set about dissecting Hawea into a crosshatch of foot trails. Every forty meters of trail they stopped to place a bait station, each amounting to a fifteen-inch length of plastic drain pipe, anchored in place by hoops of fencing wire. They allowed three weeks for the most phobic of the rats to get comfortable with these foreign objects intruding on their territories, then laid the baits. On April 10, into each of the seventy-three plastic tunnels went two wax briquettes laced with brodifacoum at five parts per thousand. Seven days later not a rat could be seen on Hawea.

With Hawea's rats now history, Taylor and Thomas raised their sights to the ultimate target. Breaksea, to their minds, was a simple matter of scaling up their proven technique by the proper order of magnitude. But for those sitting in the administrator chairs, clearing Hawea's twenty-two acres was one thing; clearing Breaksea, at two hundred times the size, with forested mountains and cliffs, was another. Breaksea, the bureaucrats scoffed, would amount to nothing but a waste of personnel and money.

Their bosses would slash their budget, but by then Taylor and Thomas had all but set sail. They had already finagled a donation from a major manufacturer of brodifacoum. They had moles in the national park system organizing supplies, a hut, and helicopter time. They had already enlisted a league of volunteers from around the world, eager to be part of conservation history. With money and manpower already in place, the department heads found themselves politically hog-tied. They would concede to the minor mutiny, but not without one parting shot over the bow.

Taylor and Thomas, in earlier interrogations, had offhandedly estimated they could finish the job on Breaksea in about three weeks. Their rough estimate now became law: Taylor and Thomas and every last one of their troops were to evacuate Breaksea Island within twenty-one days of laying poison, rats gone or not.

As they had on Hawea, they would first quell the rats' anxieties and suspicions. Weeks before the first poison was laid, they laid their trails and plastic tunnels, freely inviting all comings and goings, laying a foundation of trust. By the end of April 1988, Breaksea had a bait station within sniffing distance of every rat on the island.

On May 25 the rat team assembled once again on Breaksea for final instructions. Taylor and Thomas had a team of six volunteer rat busters, a stockpile of 770 pounds of poison, and a helicopter pilot standing by for special duty. They gathered at command central—a rustic hut with a table, some chairs, and a map hanging from the wall. The map was pierced by 744 blue pushpins marking every bait station. On the table lay a box of red pushpins. As the rats began taking baits, the blue pushpins were to be replaced by red.

The next day the Breaksea troops headed out, shouldering sacks of bait. By the end of the day 744 stations, covering every rat territory on Breaksea plus two rocky spires offshore, contained two brodifacoum-laced briquettes of wax. There would be no place to run or swim for any rat of Breaksea.

Word spread quickly among the rat community: The strangers have come bearing food, and it is good. On the morning of day two the first checks of the stations would find baits missing. By day three every station was being pilfered. The map at headquarters flushed from blue pins to red. By day five the baits were disappearing as fast as the troops could lay them.

As the trap unfolded, Taylor took to hiding himself in the bushes and observing. He soon observed that the spying game worked both ways. One particular rat seemed oddly hesitant about entering the bait tunnel. Taylor sat and sat, waiting for the rat to make its move. Tiring of the game of patience Taylor looked into his lunch bag. As he looked up, the rat was exiting the tunnel with bait.

Taylor pretended to look again into his bag, this time spying out of the corner of his eye. As he reached in, the rat darted for the bait. Taylor now knew what it was to be under the scope, with a rat dissecting his every move.

The Breaksea eradication was going off like clockwork. The big males were laying claims to the stations, eating their fill, stocking their larders, chasing all comers, and eating again. The rats could hardly wait for their next visit from the men with satchels. Taylor would walk up to a tunnel, tap it as if leaving another offering, and pretend to leave. Almost before he was out of sight, the attending rat would be in the tunnel. Taylor watched one rat making off with no less than twenty-two baits.

Despite their eons of learned distrust, the rats had badly misjudged these strangers bearing gifts. As the poison settled into their livers, their lifeblood began to leak. As the leaders lay dying in their burrows, their unguarded territories came under siege by the next-boldest cohort. And on rolled the waves of death.

Back at headquarters, the battle was being played out in two dimensions on the wall. As the massacre progressed, the map changed color again. Red pins, signifying baits still being stolen, began to revert back to the quiet symbol of blue, now signifying death. By day ten half the stations on the island had been vacated. The masses were falling, the rat tide had begun to ebb.

As the deadline approached, the map tenders grew ever more busy pulling red pins, inserting the blue. On day twenty, amid a sea of blue, one lonely little red pin remained. And this, so far as Taylor and Thomas were concerned, was the last gasp of the living dead. Having already ingested a lethal dose several times over, the last rat of Breaksea was stockpiling its own grave.

## Chapter 7

## BAJA CATS

THE FOLLOWING DAY, Rowley Taylor and Bruce Thomas were packing up and leaving Breaksea Island as planned. Three weeks to the day after laying the first bait, they had finished a campaign that the naysayers had bet would fail. As a parting act of insurance, they left at each station a double dose of poison and a single apple. They would return the next month and periodically thereafter to check for signs of Breaksea's rebirth, finding many; and to check the apples for gnawings, finding none. And indeed there would never be another rat found there.

Breaksea had reconfigured the horizons. It had demonstrated, on a beggar's budget and on hostile terrain, that a calculated dose of Kiwi audacity could beat the toughest, most intractable enemy of New Zealand's natural heritage. The gray curtain that had closed on the Wellington rat conference twelve years before, conceding a world forever compromised by invaders, had begun to lift.

With Breaksea in the bag, and their bait-station strategy now proven, Taylor and Thomas and their colleagues began eyeing bigger islands. And always with economy in mind. If an island was too big for the single storming of a little crew, they would whittle it down to size, parceling its overwhelming enormities into bite-size battlefields, deploying their bait stations in overlapping waves, one after the other, shore to shore—a rolling front, they would call it. By 1990 nearly forty of New Zealand's rat islands had been cleared by hand. That year the emboldened Kiwis added the helicopter to their arsenal, raining poison from the air and sending New Zealand's rat-free acreage on skyward trajectories.

Out of hard luck and desperation, the defenders of New Zealand's natives were pioneering a take-no-prisoners approach to biological conservation, saving island life by means of systematic eradication. And as it happened, would-be island saviors a world beyond were ripe to join the revolution.

## CLIPPERTON

In the spring of 1989 a biology student named Bernie Tershy landed a job observing seabirds from a government ship cruising off the west coast of Mexico. The project leader had arranged for a stop at a lonely little island far out at sea. The island, named Clipperton, came with masses of seabirds and an interesting history of their occupation.

Clipperton was a doughnut of nearly treeless land less than one square mile small and nearly eight hundred miles seaward from Acapulco. Sailors who began visiting the island in the 1700s were struck by this otherwise barren atoll so incongruously crawling with land crabs and stippled with nesting seabirds. The most conspicuous of the birds were two species of boobies—large, elegant high divers of the gannet family—nesting on Clipperton by the tens of thousands. The masked boobies of Clipperton constituted the species' largest colony anywhere.

Over the following two centuries the island underwent various exploitations and attempted occupations by several nationalities and one barnyard animal. American, Mexican, and French citizens came and went, but their pigs remained. Abandoned to their fates, the pigs made do by crunching their way through the scuttling fleets of land crabs and vacuuming up eggs and chicks from the seabird colonies. By the time biologists arrived in 1958 to survey the life of Clipperton, the crabs were all but gone, and the multitudes of boobies had been reduced to hundreds, the last of them huddled on a pig-free rock in the middle of the island.

It so happened that a man who had come to count their numbers—an American ornithologist named Ken Stager, from the Natural History Museum of Los Angeles County—had come toting a shotgun to bag a few bird specimens for his museum. Stager saw what the pigs were doing to the birds, took it upon himself to remedy the situation, and turned his bird-collecting tool on them. He circled the island, counting fifty-eight pigs along the way, and shot every last one of them dead.

By the time Tershy came to visit, the boobies were well on their way back, again amassing by the tens of thousands. (By 2001 surveyors would find Clipperton Island crawling under a living pavement of land crabs and inhabited by more than one hundred thousand boobies.) For him there was an epiphany in the little saga of Clipperton. Like so many of the conservation persuasion, Tershy in his fledging career had already tasted the creeping cynicism that invariably accompanied the losses. One could not live in coastal California, as he did, without witnessing the daily retreat of the wild places. Nor could one spend any appreciable time studying seabirds, the vocation Tershy was now courting, without repeatedly finding colonies decimated and littered with half-eaten carcasses.

But here, in this simple story of Clipperton, was an inspiring revision of the tired and tragic ending. A lone man with a gun had restored a hundred thousand seabirds—had revived a moribund island ecosystem. Formerly resigned to the narrative of doom, Tershy suddenly envisioned himself a savior of species.

Tershy's closest colleague and sounding board was a fellow seabird biologist working in La Paz, on the southern coast of the Baja peninsula. Don Croll had studied seabirds across both hemispheres, from penguins in the Antarctic to murres in the Arctic. He and Tershy had both witnessed the same recurring carnage, of seabirds congregating so faithfully and fatally in an island's broken seclusion. But on the flipside of that fatal flaw lay opportunity: Vanquish the threat, restore the masses. And the returns would multiply far beyond seabirds.

Acre for acre there was no real estate in the world of endangered life-forms more precious than that of the ocean's islands. Comprising 5 percent of Earth's landmass, islands had come to harbor one in every five species of bird, mammal, and reptile. They had also shouldered 63 percent of all their extinctions recorded during human history. Island species had come to populate nearly half of the list of the world's endangered. And most of those owed their endangerment to invaders.

For a pair of conservation biologists wanting quick and massive impact, there was no calling more obvious than making the islands safe from invaders. And for Croll and Tershy, there was no better place to start than Baja.

## DESERT ISLANDS

Eastward between Croll's university on the Baja peninsula and the Mexican mainland lay the Sea of Cortez. Thirty miles west lay the Pacific Ocean. Within these coastal waters of Baja arose some 250 desert islands, generally distinguished by a scarcity of human settlements and a striking profundity of wildlife and wilderness.

These were islands stark in appearance and astonishingly rich in their life-forms. On the sun-beaten beaches and sea cliffs, great colonies of seabirds would gather to nest. The isolation that drew the flocks had also modeled the islands' resident individuals into peculiar forms—of sparrows and rodents, lizards and snakes, changing form from one island to the next. For modern-day wannabe Darwins, this was the magic kingdom.

For the biologist concerned with saving the creations, however, the Mexican islands more resembled an inner-city emergency room at midnight. The islands came plagued by the usual cast of suspects from the mainland, the rogues' gallery of burros, goats, rabbits, rats, and cats—grazers, browsers, meat hunters, and egg thieves, nibbling and gnawing and mauling their way through the evolutionary oddities and seabird masses. The list of casualties ran long. Recently gone were the Todos Santos rufous-crowned sparrow and the Todos Santos wood rat. Gone too were the San Roque white-footed mouse, the McGregor house finch, and the Guadalupe storm petrel. Nineteen species of native animals had been extinguished from the Mexican islands since the time of human settlement, eighteen of them with help from the modern menagerie of invaders.

Croll and Tershy, as would-be island conservationists, found they weren't alone. Beyond Ken Stager's impromptu pig eradication on Clipperton, there was the ongoing campaign of the U.S. Fish and Wildlife Service, already years into liberating Aleutian seabirds from imported foxes. A team of independent biologists were even then readying themselves to eradicate bird-killing cats from Wake Atoll, in the western Pacific. And leading the way from Down Under were those pioneering Kiwis, escalating the offensive against the raiders of their island nation. Croll and Tershy believed they could do likewise for Baja.

Though the two soon migrated to faculty positions at the University of California, Santa Cruz, their focus remained fixed on saving the Mexican islands. They teamed with José Ángel Sánchez-Pacheco, a marine biologist from Baja with a kindred passion for island life and a behind-the-scenes talent for navigating the Mexican bureaucracy. Their plan took shape. They would seek their funders in the richer pockets north of the border, rally their Mexican colleagues south of it, and together, as an international team, save this beleaguered part of their world under a bilingual banner of island conservation.

If they could somehow deal with the cats. The two most immediately troublesome invaders blocking their way were Baja's rats and cats. With the blood-thinning brodifacoum on the shelves and the New Zealanders' bait-station protocol now on the books, the conservationists already had the means and the model for confronting the rats. The cats, on the other hand, had yet to reveal any such silver bullet weakness. And as of 1994 there were twenty-six islands in northwest Mexico harboring menacing populations of them. The trio of conservation biologists found themselves woefully out of their league as cat hunters. Hence their first and most important hunt sent them in search of one.

## THE CATMAN

Bill Wood was a retired trapper from the desert flanks of California's southern Sierra Nevada. Wood's specialty was the bobcat, a brawnier, stub-tailed cousin to the domestic cat. Over the years, Wood had honed his craft, earning a comfortable living trapping bobcats and selling their pelts. He had gained a reputation in Southwestern trapping circles as the king of bobcats. To some he had even come to embody his quarry, wiry and alert, right down to his probing feline eyebrows.

The recipe to Wood's success amounted to several decades of trial and error, and a dash of the intangible better described as art. Wood shunned gimmicks. His trap of choice was a standard Number 3 Victor leghold trap, featuring a pair of spring-loaded steel jaws triggered to bite upon the weighting of a cat's foot. Beyond that, he parted company with the crowd. Where others baited their traps with veritable billboards of elaborate lures and witches' brews of foul-smelling scents, Wood placed his faith in the subtle skill of placement. He had learned to read the land for its natural corridors and travel ways, the most likely paths by which the cat would cover its territory. If need be, he would delicately arrange surrounding rocks and bushes into chutes to guide the unsuspecting forefoot exactly to the center of his trap. Precision was key. An inch or two to either side risked a misfiring or an escape. Wood didn't lose many cats to miscalculation.

Or to the competition. Wood once found himself sharing a trapping territory with a man who had all but mined the mountain with traps. Wood laid out his three traps to the other man's seventy-five. And every day on his rounds he would stop by the enemy's camp for coffee, to listen humbly to yet another sad tale of rotten luck. The man eventually packed his gear and left, empty-handed and grumbling something about a scarcity of bobcats, unaware of the daily stash of pelts piling up in the back of Wood's truck.

Wood went to great lengths to guard his secrets. If he erred so much as to make a track in the sand, he would brush it out, to keep both bobcats and snooping trappers clueless of his presence. He had no intention of letting freeloading neophytes go "to school on him," stealing with a glance the techniques he'd honed over years of dirt time. One trapper offered Wood a thousand dollars just to let him tag along. "Didn't seem like a good deal to me," Wood would later say. "I was making that much in a day."

Croll and Tershy caught wind of Wood's ways with wild cats and called for his help. He was by then enjoying the retired life with his wife, Darlene, remodeling one of his several vacation homes, trapping as it pleased him, and indulging an addiction to fishing. But he was barely sixty years old and still harboring the adventurer's itch. Now here was this stranger on the phone, offering to pay him for a stint on a sunny desert island in Mexico. And oh, by the way, the fishing was great. Wood, who spoke no Spanish and knew next to nothing of trapping feral cats, said what the hell, packed his gear, and headed south.

Before sailing off on his Baja assignment, Wood, the hired catman, was briefed on the adversary. The feral cat was the vagabond version of the world's most popular pet. Its alter ego, the domestic cat, had descended from *Felis silvestris*, the wildcat, a lithe little hunter originating in the Near East. It was there, some ten thousand years before, in the rich Mesopotamian river valleys of Iraq and Israel, that the first cultivated crops of wheat and barley began to replace the wild produce of hunting and gathering as the human's basic sustenance. The prevailing speculations picture a pioneering wild cat wandering into one of the rudimentary villages of the dawning age of agriculture, drawn by the rats and mice foraging in the first farmer's new grain bins. The farmer, quickly tiring of the pilfering rodents, grew to appreciate these wild little predators lurking about. And when the resident she-cat eventually had kittens, the kittens eventually became pets. It was a bargain for both. The cat got her mouse, the farmer got her grain, and both even came to enjoy each other's company on the side. In the shrewdest of evolutionary plays, the wild cat domesticated itself unto the wealthiest provider in the animal kingdom.

Over the next nine thousand years, progeny from that domestic cat—and perhaps from a handful of others hitting independently upon the same winning ticket—would come to number more than half a billion. More so than the appeasement-driven dog, the domestic cat retained its wild independence, tiptoeing a fine line between purring lap cat and hissing spitfire, a certain wildness that would serve it well whenever the interspecies marriage went sour. Cats could be abandoned just about anywhere and make do. They sailed the Atlantic with Vikings, sailed the Pacific with Captain Cook, explored the subantarctic with sealers. They were commonly presented as goodwill gifts to native islanders. Other times they jumped ship and made themselves at home. The cat rode the coattails of humans to the ends of the earth, and many islands in between.

The cat's global wanderings were to be exceeded only by the rat's. But the speed with which the cat could ravage island faunas was exceeded by none. In the hierarchy of Oceania's invading predators, the cat came to be classified as a superpredator. Its mouth was a butcher's array of artery-slicing canines and meat-shearing molars, its paws concealing twenty switchblades. The weaponry was directed at times by a brain hardwired and hair-triggered for repeated attack regardless of hunger. Such was the nature of the beast behind the gruesome slaughters so commonly greeting the seabird biologist, the dead lying en masse with broken necks and missing brains and hardly a feather otherwise ruffled.

Among grounded gatherings of nesting birds, the litany of cat carnage ran long. On the New Zealand island of Raoul, a rookery of sooty terns likely numbering hundreds of thousands when the first cats were put ashore in the 1800s was gone by the 1990s. In the Kerguelen Archipelago, a subantarctic wilderness of penguins and petrels and albatross, cats at the peak of their killing were calculated to be removing nearly one and a quarter million seabirds per year. On Ascension Island, in the middle of the Atlantic Ocean, cats delivered by European colonists would by the twentieth century drive a gathering of some twenty million seabirds to within 2 percent of annihilation, the survivors left clinging to the sheerest cliffs and offshore sea stacks.

By the time Bill Wood arrived in Baja, feral cats of the world had extinguished at least thirty-three bird species and decimated uncounted others. And birds were only the more obvious victims. In the era of the feral cat the Mexican islands had thus far lost more than ten of their unique rodents. Distinctive forms of lizards and snakes had all grown scarce or disappeared with the coming of the cat. This was the buzzsaw of biological diversity that Wood was hired to destroy.

For Wood the transition from Sierra Nevada bobcats to Baja house cats entailed certain adjustments, the least of which involved retooling for the new quarry. Never mind that these stray cats were animals a third the size of his specialty, demanding smaller traps, lighter triggers, and more delicate placements. For Wood the ultimate challenge was maintaining his coveted independence.

Tershy had started Wood off, handing him the latest scientific papers on the subject, prescribing orderly grids of traps, mathematically configured to cover every square inch of cat territory. Wood secretly put them aside. His tool of choice was the scalpel, not the sledgehammer. As Wood went to work, meticulously setting his traps, Tershy would sometimes follow, camera in hand, documenting every detail. Wood stifled a desire to throw Tershy and his damned camera in the ocean. He could imagine his trapping secrets, so painstakingly earned and rabidly guarded all these years, splashed on a big screen before a packed auditorium. What Wood could not imagine then was that one day the man spilling it all at the podium would be he.

## AMBASSADOR BILL

Much as he might have wished otherwise, there were feral cats enough in Baja to defeat even the single greatest trapper in the land, and only one Bill Wood to go around. If Wood was going to finish his job, he was going to need help. And others were going to need his.

Wood began appearing before Tershy and Croll's Mexican counterparts like some strange sort of shaman from the North, sent to cure their curse of cats. The Mexicans were primarily shooters, hunting by night with rifles and headlamps. Few seemed particularly interested in learning Wood's mysterious leghold weaponry. Safe enough with his secrets, he would lay the lines, setting the traps in his inimitable way, and with the help of an interpreter would instruct the others to check them. Then on to the next island he went.

As Wood toured the islands, the purpose of it all began to sink in. The stakes in Baja figured not in the profit of bobcat pelts but in million-year lines of evolution. These were no longer fishing trips disguised as work. This was a noble cause he had stumbled into. Applying his singular skills to saving wildlife gave Wood a satisfaction he'd never known as a commercial fur trapper. "It got to where I realized I'd done a lot of damage catching and hunting things," he would later say. "To give something back—that was important to me."

Wood, the iconoclast bobcat trapper turned grassroots extinction warrior, began recruiting troops to his new mission.

He looked for outdoor sorts requiring no handholding—hunters, fishers, people of the land. Experience with traps was optional. (Wood figured it would take more work to untrain bad habits than to teach the right ones from scratch.) Allowing that there would still be night shooting to bag trap-shy cats, Wood looked for those handy too with the rifle and spotlight. And allowing that such night hunting was illegal in Mexico, the résumé Wood was seeking fairly well described the poacher.

Which fairly well described Miguel Angel Hermosillo. Hermosillo was working as a crew member of an eradication attempt already under way, directed by Tershy and Croll's Mexican partners from the Universidad Nacional Autónoma de México. He was also a part-time worker with a Mexican railroad company. His skills also involved shooting deer for money and for meat, under cover of darkness and out of earshot of the law. Unlike others Wood would attempt to train, Hermosillo grasped not only the intricacies of Wood's technique but their purpose. Hermosillo was no longer hunting for money but for the cause.

Wood took Hermosillo under his wing, taught him his techniques and trapping sets, trusted him to run the lines in his absence. Hermosillo spoke no English. Wood spoke no Spanish. Wood spoke to Hermosillo using his Spanish-English dictionary and a stick, drawing pictures in the dirt. The two understood each other perfectly. When Wood reported back, the first thing he recommended to his bosses was that they hire Hermosillo. The first employees of Croll and Tershy's fledgling island campaign thus comprised a retired bobcat trapper from California and, now, an itinerant deer poacher from Mexico.

And so, Wood began adding to his little ragtag battalion of island conservationists. By night the hunters would lure their quarry, squeaking like mice and wailing like wounded rabbits, aiming their headlamps and .22 rifles into the cat's incandescent green eye-shine. By day they checked their traps and ran the dogs. Jack Russell terriers—two-thousand-dollar dogs from elite hunting lines that had proved their mettle on U.S. mountain lions—joined the teams, tracking Baja cats and rabbits and flushing them into the gunners' sights.

As Wood made the rounds, his reputation transcended from that of master trapper to that of Baja's patron saint of island fauna—St. Francis with a twist. It turned out that Wood the cat whisperer had a magical way with people too. Wherever he stopped, he would pull out his little book of photos, filled not only with telling scenes of the feral cat's seabird carnage but also with images from back home—pictures of animals he'd hunted, fish he'd caught. With his pointing finger and phrase book Spanish he charmed all hosts. By his second visit to Baja, Wood never again needed to book a motel or hail a taxi. His new best friends wouldn't think of him staying anywhere but in their homes.

And if Wood wasn't ambassador enough, he came with his sidekick Freckles. Early on Wood had picked up the little white terrier as his hunting companion. Freckles was a bit of a runt, the last pick from an otherwise star-studded litter. Her grandfather was none other than Wishbone, of TV fame, a talking, daydreaming, well-versed-in-classic-literature kind of dog. Wishbone's undistinguished little granddaughter Freckles overcompensated for her size with a tenacious talent rounding up island cats and rabbits, and a star power that threatened to steal her master's stage. Tourists cruising the islands would occasionally spot the now-legendary Bill Wood on his rounds and come running, gripped in celebrity fever. "Where's Freckles?" they would yell.

## SURFBOARDS AND BIRKENSTOCKS

Within two years of Bill Wood's arrival, Croll and Tershy's team had gathered two more linchpins to their Baja campaign. Brad Keitt had come to the team as a young college graduate and old buddy of Tershy's, with the résumé of an itinerant wind-surfer, kayaker, eco-tour guide, and fellow seabird junkie. When Tershy offered Keitt a project in need of a student—involving a mysterious ocean-wandering seabird named the black-vented shearwater, nesting almost exclusively on a little island called Natividad (which also happened to feature a world-class surf break)—Keitt's decision wasn't one of yes or no, but whether to pinch himself.

Keitt's academic questions concerned the natural history of a bird hardly known to science, and in danger of remaining that way forever. The black-vented shearwater bred nowhere else but the northern coast of Baja. More precisely, 95 percent of the world's black-vented shearwaters gathered each season on the single island of Natividad, an island now swarming with cats.

The original sources of those cats—as hoped-for hunters of mice—were the four hundred or so people of Natividad's fish-ing community. The island's fishermen had come to consider the shearwaters as just another kind of *nocturno*, a generic term lumping together all burrowing seabirds of the night. Keitt gave them another way to think of the shearwaters. He shared with them his discoveries, of a bird with wings that could fly it all the way to Canadian latitudes, then propel it to incredible depths of the sea in pursuit of little fish. He would relay such tidbits to those making their living diving for abalone and lobster and watch for that aha moment when the particulars hit home. "You mean these birds dive deeper than we go in our scuba gear?"

Keitt took school groups on field trips to the shearwater colony, snaking fiber-optic probes into the birds' burrows, amazing the kids with live videos of shearwater chicks never before imagined. It was a secret village of birds, hidden all this time, right beneath their feet.

He convinced Natividadans young and old that this amazing and beleaguered bird living in their midst was *their* bird. Isla Natividad was soon sprouting billboards and T-shirts emblazoned with images of the black-vented shearwater. The school and its soccer team had a new mascot.

When Keitt then showed the villagers what the feral cats were doing to their shearwaters, the massacres amounting to one hundred birds per week, the islanders didn't just agree that the cats should go; they demanded it.

Keitt's arrival at island conservation was soon followed by that of Josh Donlan, a long-haired, footloose biology major out of Virginia's James Madison University. Donlan had capped his college career with a twenty-year moment, deciding that before any grad school or meaningful employment he would sample the good life. After crossing the United States, adventuring all the way to Alaska, he eventually found himself paddling a kayak the length of Baja, on a four-month tour of the Gulf of California. He was drawn to the islands, putting ashore and exploring their miniature universes of evolution. For a young biologist looking to make sense of an otherwise complicated world, this was the holy land. But for all the seductive rawness of Baja, it was one decidedly unromantic evening on an island called San Jorge that finally hooked him to the place. On his first night on San Jorge, Donlan awakened to the scurrying of little feet and the thumps of little bodies jumping against the walls of his tent. There went the holy land.

As Donlan continued his tour of the islands, the pattern

repeated. It became more the rule than the exception to find the Baja wilderness running with urban rats and feral cats, its fabled seabird colonies littered with carcasses. And if it wasn't the rats and cats eating Baja alive, it was crews of donkeys, goats, and rabbits mowing the delicate island flora to nubs. This was a paradise in need of rescue.

Donlan, like Keitt, was soon thereafter enrolled in the graduate program at UC Santa Cruz, ostensibly as a master's student of biology under Tershy and Croll, in practice a new recruit to their eclectic squad of island saviors. Donlan declared that his first official order of business was to return to his baptismal shores of San Jorge and rid that island of rats. He loaded up his truck, snuck a load of brodifacoum across the Mexican border, enlisted a crew of local fishermen to help spread it, and within the year had made the island safe again for its seabirds. For the twenty-five-year-old aspiring biologist in Birkenstocks, having enlisted in conservation's losing war against extinction, it was a heady initiation to the front lines. With a truck full of poison, a few fishermen, and no fanfare, Donlan had rescued an island.

And so it came together, this unlikely coterie of conservationists—a veritable motley crew of egghead academics, professional poachers, ex-hippies, trappers, Latinos, and Yanks—allied in the cause of island species. The team officially became the Island Conservation and Ecology Group, mirrored in name and mission by their Mexico comrades, Grupo de Ecología y Conservación de Islas. Bankrolled by shoestring budgets and unjaded optimism, the island conservationistas started clearing islands of their animal threats, one after the other.

The crews would one day look back fondly on those halcyon days of misadventure among the Mexican islands, of runaway hunting dogs and run-ins with drug runners, of tense moments at the border, their trucks loaded with guns and ammo. There would be breakdowns in the middle of the desert, breakdowns in the middle of the sea, little vessels full of island crusaders with their little dead engine in pieces on the floor, drifting helplessly toward the black horizons of the Pacific. There was an inadvertent collision with a whale. The prevailing philosophy for coping with the daily dangers was epitomized by Wood's wife, Darlene. "She didn't say too much," said Wood. "She just raised my life insurance."

Within five years of their opening salvos in Baja, the cross-border team of island conservationists had cleared invaders from nine islands. They'd eradicated rats from San Roque and Rasa islands, rabbits from Natividad, and goats and burros from two of the San Benito Islands. They'd eliminated cats from the islands of Isabel, Asuncion, Coronado Norte, San Roque, and Todos Santos Sur, with those harassing the shearwater haven of Natividad soon to follow.

In an age of endless bureaucratic delays and budgetary over-runs, the little crew of academics from Santa Cruz and their unlikely allies from the rowdier side of the tracks would eventually protect eighty-eight species of Baja's singular fauna, along with 201 seabird colonies. For less than fifty thousand dollars per rescue, the island saviors had quietly engineered one of the most potent and streamlined campaigns for conservation ever imagined.

Yet those halcyon days of anonymity were not to last. With success came stature, and a call for help from a besieged island called Anacapa, an island lying within sight of Los Angeles.

*Chapter 8*

# ANACAPA

IN 1853, TWELVE miles off the coast of Southern California, the paddle steamer *Winfield Scott* wrecked upon the rocky
shores of Anacapa Island. In the year 1990 an oil tanker named
the *American Trader* had an unrelated misfortune, running over
its anchor off the coast of Huntington Beach. Separated by more
than a hundred years and nearly as many miles, the two events
would eventually converge to trigger a landmark battle in U.S.
conservation history.

The *American Trader*'s punctured tanks disgorged four hundred
thousand gallons of Alaska crude, which would very publicly kill
some thirty-four hundred seabirds. Of the thirteen-million-
dollar settlement the court imposed on the tanker's owners,
one and a half million was to go toward undoing some of the
damages.

The most worrisome mortalities from the spill had been borne
by the Xantus's murrelet, a robin-size seabird numbering just a
few thousand breeding pairs short of extinction. One obvious
way for the offenders to make amends was to make someplace
safe again for the murrelets to breed. And one obvious place to
do so was Anacapa.

Anacapa is three slips of jagged rock in the archipelago comprising Channel Islands National Park. Bearing a footprint of little more than one square mile, Anacapa's physical dimensions belie its biological capacity. Its razor-edged cliffs, pocked with caves and crevices and surrounded by forbidding moats of seawater, constitute what would otherwise be prime real estate for nesting seabirds. If not for the rats.

The long-forgotten wreck of the *Winfield Scott*, like that of the *American Trader*, also disgorged a bit of cargo, albeit in the form of a stowaway rat or two. The results of that little uncelebrated landing on Anacapa—adding more strange blood to an island already beset by immigrant cats, sheep, and rabbits— would dwarf the *Trader*'s oil-slick body count.

By the end of the twentieth century the black rats of Anacapa had decimated the Xantus's murrelet in one of its final refuges. Moreover, the whole of Anacapa's specialized biota had apparently fallen under the rodents' giant shadow. Crabs and urchins on the beaches, lizards, grasshoppers, and wildflowers on the hills and headlands—all had become easy prey.

The rats of Anacapa had more intimately impressed themselves upon East Anacapa's park rangers and tourists as the ubiquitous little gremlins forever breaking into foodstuffs and camping gear. Park workers had tried fending off the rats with a few snap traps and the poison warfarin. Their weaponry and resolve proved too weak. They invariably left the wilder reaches of Anacapa unguarded, leaving an eternal wellspring of rats to spill forth wherever the coast was clear. Time went on, the rats kept coming, money ran short, and the white flag went up.

With the serendipitous spilling of the *American Trader*'s oil, and the windfall of mitigation money that washed ashore in its wake, the National Park Service got serious about taking Anacapa back. It sent for help from a troop who'd recently been making a reputation for themselves on similar terrain south of the border, in Baja, calling themselves the Island Conservation and Ecology Group.

## THE BELLY OF THE BEAST

By now it was clear to both the Park Service and the band of island conservationists from Santa Cruz that to merely control the rats was to forever bail the leaking boat. The only permanent fix for Anacapa's rat problem was, to their minds, eradication. And the only practical means of accomplishing that was the potent blood-thinning poison brodifacoum, broadcast by helicopter and blanketing Anacapa shore to shore.

The aerial technique had begun working wonders down under in New Zealand, where the Kiwis were clearing rats from islands tens of times larger and hundreds of times more remote than Anacapa. But this was the United States, land of legal hurdles, bureaucratic hoops, and litigation. Its citizenry had grown suspicious of those professing to perform noble acts of public service while raining dangerous chemicals upon their heads. Their society had been rudely awakened by Rachel Carson's 1962 classic exposé, *Silent Spring*, which raised the specter of a world blithely poisoned by the pesticide industry—of a toxic world where no birds sang, where cancers ran wild, and where the well-being of humanity itself lay in doubt.

Americans had become suspicious as well of their own government's ongoing war against wildlife. Troops of tax-funded trappers, shooters, gassers, and poisoners had been slaughtering American wildlife by the millions every year for nearly a century. Operating under the official sanctions of the U.S. agency Animal Damage Control (later rebranded as Wildlife Services), the exterminators went after blackbirds, prairie dogs, coyotes, cougars, and wolves and a host of similarly pigeonholed vermin across the country. Too often they killed indiscriminately, too often to the demise of innocent bystanders, and too often arrogantly, in the face of repeated rebukes from scientists and humanitarians. It had become all too easy to lump such government-sanctioned fouling of American soils and wanton slaughters of wildlife with the island saviors' ironic mission of killing for conservation. Anacapa would of course be questioned. To think of dropping truckloads of poison on a national park within sight of the sixteen-million-person megalopolis of Los Angeles was to aim a slingshot at a hornet's nest.

The Anacapa campaign was to be led by a newcomer to the Island Conservation crew, a rising expert on brodifacoum, the predominant anti-rat weapon of the day. Gregg Howald had conducted his graduate research on Langara Island, off the coast of British Columbia, probing for chinks in brodifacoum's solid record. Langara's eradication history had been triggered by the arrival of ship-jumping rats, which over the decades had reduced a world-class colony of some two hundred thousand ancient murrelets to less than fifteen thousand. With the advent of brodifacoum and Rowley Taylor's bait-station protocol, in 1995 the rats were attacked in turn. Langara, measuring thirteen square miles, was a project nearly twenty times the size of Taylor's battle for Breaksea, and more than ten times the size of Howald's Anacapa campaign to come. But the issue raised by Howald's investigation created concerns beyond the mere upping of acreage.

Among the dead of Langara, Howald had uncovered a small but especially worrisome number of rats lying exposed to the scavengers—rats that were supposed to have died in their dens. One rat was found hanging thirty feet high in a hemlock, its hindquarters and internal organs ominously missing. Testing their suspicions, Howald's team captured a sample of bald eagles. Each had brodifacoum in its blood. So too did the thirteen dead ravens they came upon. One dead raven appeared to have been scavenged in turn. Lying dead beside it was a bald eagle.

The Langara eradicators went on to accomplish their primary objective with yet another record-smashing performance, all thirteen square miles cleared of immigrant rats in less than four weeks. But Howald's work had added an unofficial asterisk to the record book. "We conclude that there is a very real risk to some avian predators and scavengers," reported Howald and colleagues, "from both primary and secondary exposure to brodifacoum when used to remove rats from seabird colonies in the Pacific Northwest environment."

## SHELTER FROM THE STORM

The cautions from Langara didn't bode particularly well for those now looking to remove rats from seabird colonies in the Southern California environment of Anacapa. Anacapa came with big potential for collateral casualties, the biggest in the form of a mouse. The Anacapa deer mouse was a Channel Islands native, a Dumbo-eared, doe-eyed mouse with an endearing touch of island tameness. Its DNA profile set it slightly but significantly apart from all other deer mice, but unfortunately for the purposes of the eradication crews it was still at core a rodent, a chisel-toothed opportunist every bit as vulnerable as every rat they were out to kill.

Not to mention what an island littered with poisoned rodents might mean for Anacapa's native suite of raptors—its barn owls and burrowing owls, its red-tailed hawks, American kestrels, and peregrine falcons—none of them necessarily averse to taking advantage of a dead or dying rat, all of them rigidly protected by law.

In a world of islands under siege by invaders, Anacapa was not by a long shot the most physically intimidating to defend. But on the scales of biological complexity and in the courts of public opinion, it loomed enormous.

The planning alone promised years of work. There would first be an environmental impact statement and reams of permits and paperwork explaining in numbing detail this atypical conservation proposal. If all was approved, the poison was to drop from an industrial grain hopper slung beneath a Bell 206 helicopter, showering bait with computer precision and leaving no rat-worthy patches of refuge on Anacapa unpoisoned. That was the easy part.

For the natives sure to be caught in the firestorm, there would need to be a separate operation. The island mice were to be gathered by the hundreds and caged out of harm's way until the poison had cleared. The raptors too would be trapped and sequestered, or released on the mainland. As a final layer of fail-safe, the poisoning of Anacapa's three islets was to be staggered a year apart, allowing recovery of one before the next was bombed.

Such was the plan eventually approved. It was all very avant-garde, employing the latest special weapons and tactics of eradication technology, melded with endangered species husbandry and a hint of high-stakes shell game. It was, after all, the continent's first-ever rat poisoning to be attempted by air. And that, for better or worse, would attract attention from more than admirers.

## SABOTAGE

On October 24, 2001, a week before the first bait was scheduled to hit the ground, field crews conducting last-minute checks spotted two men landing their inflatable dinghy on a beach on East Anacapa. Through binoculars they watched one of them reaching into his backpack and flinging something, as if tossing a Frisbee. Later that evening, when the men found themselves stranded by a broken motor, the Coast Guard paid a visit and took names. Word got back to Park Service headquarters, and the phones started lighting up: *Puddicombe's on the island!*

Rob Puddicombe—whose varied résumé included stints as a commercial diver, a bus driver, a volunteer wildlife rehabilitator, and, more important of late, an outspoken critic in the local papers of the impending poisoning of Anacapa—was by then a familiar name to the eradication team. When Howald heard that his crew had seen either Puddicombe or his accomplice throwing something, the toxicologist had an immediate hunch: "Go back and tell me if you find any pellets."

Howald's crew found the ground littered with rat kibble, almost identical in appearance to the poison bait awaiting deployment. Lab tests came back confirming his suspicions. The pellets had been infused with vitamin K, a standard remedy for brodifacoum poisoning. "My god," thought Howald, "these guys are attempting to spread the antidote."

Puddicombe soon had allies. Five days later the Park Service received notice from the Fund for Animals, a national animal rights organization based in New York, that the Fund and Puddicombe intended to sue. They objected to the killing, to the poisoning of the Anacapa wilderness, as "arbitrary and capricious." They questioned whether the rat was really a threat to the Xantus's murrelet.

The project ground to a halt while a federal judge deliberated. "It caught us off guard," said Kate Faulkner, the Park Service's chief of natural resources. A few years earlier Faulkner's agency had felt a similar sting of societal venom while eradicating feral pigs from the neighboring island of Santa Rosa. "We thought we had very compelling reasons for eradicating rats on Anacapa," she said. "We knew we had to do a little explaining. But we thought rats would be more acceptable than pigs."

## A Nicer Way to Die

For those out to save the last of Anacapa's murrelets, when deciding between an individual rat and an entire colony of seabirds, there was no choice. To allow a work of evolution eons in the making to be extinguished by a ubiquitous species of rat running amok was to stand complicit in a crime against nature.

"To some people it might seem kind of extreme," said murrelet biologist Darrell Whitworth, whose surveys on Anacapa had become a repeating tour of plundered nests and emptied caves. "But rats are everywhere. This is where murrelets nest. They're going to nest here, or they're not going to nest anywhere. I'd much rather see a lot of murrelets here than a lot of rats."

Not so the rat's defenders. "And who are humans to call another species invasive, huh? That is a joke," Puddicombe told a reporter from the *Washington Post*. "Species go extinct all the time. That's the philosophical difference. These animals are here and alive now. Their lives have value."

From somewhere between the poles came a more nuanced question, of conservation with compassion. Asked Marc Bekoff, professor emeritus of ecology and evolutionary biology at the University of Colorado at Boulder, "Is this the most humane way?"

Bekoff came with a rather rare and, some might say, conflicted résumé, as a noted scientist of animal behavior and an outspoken advocate for animal rights. (Early in his career he had dropped out of med school, refusing to kill the house cats required by his experiments.) "I have a really big problem with this carte blanche, overriding theory that invasives should be killed," said Bekoff. "It's become a numbers game, the argument being there are so many rats it doesn't matter if we kill some of them. It's the veil they hide behind. I can't tell you how many times I've heard, 'I'm a conservationist, this is how we do it.'"

"I'd like to get people thinking about humane alternatives," he continued. "Too many people just don't know about the studies."

The studies to which Bekoff referred were those raising new ethical questions about the conservationists' lethal new means. It seemed there was more pain and suffering for the cause than was readily appreciated, and much of it inflicted by the work-horse weapon of the rat eradicators.

Brodifacoum had been commonly assumed to be a relatively peaceful way of passing. The poison evoked the image of a woozy rat retiring to its burrow and curling up for the final sleep. Dale Kaukeinen, the man who'd ushered brodifacoum into the lime-light of rodent control, who'd watched more than his share of rats die, had come away with tentative assurances. "With internal bleeding I'm sure there is some discomfort," he said. "But I never saw rats vocalizing or thrashing about."

Some who had looked a bit closer had come away less comforted by brodifacoum's reputed kindness. Kate Littin, a physiologist and a technical adviser to the Animal Welfare Group of New Zealand's Ministry of Agriculture and Forestry, had tested brodifacoum specifically for that elusive quality of humaneness. Littin herself poisoned her rats and documented in detail their demise. Death did not always come comfortably or quickly. As the poison took hold, listless rats took to crouching, backs hunched and heads drooping. Littin watched rats lying half paralyzed for hours, pushing and pulling themselves across the floor. The rats under her observation took, on average, a week to die. Brodifacoum's delayed onset of fatal symptoms, the very quality that made it so diabolically perfect for overcoming the rats' hypersensitive danger meter, also made it one of the nastier ways to die.

Littin would later team with her colleague Georgia Mason on a broader investigation of humanity's anti-rat artillery, which, besides doing little for brodifacoum's benevolent reputation, revealed how scarce were the sympathies for a rat. Rodents were regularly being gassed with eye-searing acid, crushed by snap traps and mired in glue traps (sometimes skinning themselves in their panic to escape). Trappers would find survivors hours into their ordeals, covered in their own excrement and screaming.

Finally, there was brodifacoum and its family of anticoagulants. Brodifacoum's painful hemorrhaging and slowness of death, not to mention its tendency of killing unintended victims, left the most popular modern weapon in the war on rats flunking its humaneness exam.

"We can see that rodents are routinely subject to cruelty," Mason and Littin concluded, then adding a measured dash of diplomatic understatement. "This highlights an interesting paradox in the way we treat different animals."

Littin readily noted that there were times when humaneness might understandably be shelved for more pressing human concerns. This was, after all, a creature accused and guilty of chewing holes through homes, biting infants in cribs, and carrying disease. The rat—as germ-spreading accomplice to the Great Plague—had its vermin's image burned into humanity's historical memory. Rodents were every year intercepting as much as a third of the human world's food supply; people were going hungry for want of fewer rats. For these and a host of other, less compelling reasons, the rat owned a reputation among Western cultures as low as the scale reached. A survey of American college students in the 1970s, rating their fondness for fourteen common animals, ranked the rat dead last, trailing even such cultural favorites as the worm, the shrew, and the spider. And rodents, as the island squads were now making clear, had been abundantly demonstrated to be taking a serious bite out of the world's roster of life-forms.

But was this an animal whose corporal punishment invariably fit the crime? "For most people it's a no-brainer," said Littin. "Don't worry about poisoning. The line we take, in New Zealand and Australia as well, is rather than totally ignoring animal welfare and saying these pests aren't worth considering in that regard, at least be aware. Is there something you can use that has less impact on animal welfare?" Littin's concerns, like Bekoff's, were colored by a startling body of neuroscience revealing, in the mind of the rat, traits of a disturbingly personal nature.

## THE JOY OF RATS

One day in 1997, a psychobiologist named Jaak Panksepp walked into his lab at Bowling Green University and announced, "Let's go tickle some rats." Panksepp and his students had for years been eavesdropping on their lab rats, tuning in with ultrasonic sensors and a growing curiosity about the high-pitched chirpings emanating from young rats as they played. What they were hearing they could only describe as rat laughter. So to test, they started tickling. Panksepp and his labmates would reach in with their hands, pouncing playfully, tumbling the little rats, tickling their bellies. The acoustic sensors went ballistic. The rat children were giggling themselves silly.

To Panksepp's ear it was the gleeful shrieking of kids playing tag in the school yard. In time the rat ticklers needed only to present their hands and the little rats came running as if answering the opening bell of recess.

Panksepp's curiosity had revealed a disturbingly endearing alter ego of one of the least loved animals not named the mosquito. He had discovered in the rat an emotion once believed to be the sacred province of humans and only lately and begrudgingly bestowed upon such perennial human favorites as the dog and the chimp: Panksepp had discovered in the rat the emotion most aptly defined as joy.

And where there was joy, could such emotions as fear and anxiety, sorrow and empathy, be far away? That question would soon after be answered by the rat's little cousin, the laboratory mouse, albeit through an ironic mode of sadistic inquiry. Researchers at McGill University, in Canada, had come upon the idea of injecting a mouse with acetic acid, therewith setting the creature to writhing in burning pain. Most important, they made sure its cagemate was watching. The scientists then likewise injected the cagemate. It too writhed, but more frantically than the first, its torture magnified by the memory of having watched its companion suffer the same fate. It was one mouse feeling the other's pain. In 2006 the researchers reported their astonished observations as the first evidence of any animal beyond humans and their fellow primates showing the emotion of empathy. (No mention was made of whether those administering the pain experienced any empathy of their own.)

The Canadians' empathic mice only added to a growing canon of research rapidly closing the supposed gap between the lofty universe of human emotions and the less-exalted domain of the animal kingdom's lower ranks. As early as the 1950s caged rats had been documented deliberately forgoing food to spare their fellow rats an electric shock. When rats witnessed their neighbors being decapitated, their blood pressures soared, their hearts raced.

Inside the heads of some of the most reviled creatures on the planet were sensitive minds harboring emotional kinship to the species so blithely torturing them. So concluded the studies buttressing Littin's and Bekoff's concerns, begging questions rarely discussed in eradication circles. Death had become accepted as a necessary ingredient of the conservation prescription, the mantra among the island saviors being that the ends justified the means. Interlopers were attacked, natives rescued, mission accomplished. To spare the island invader its singular pain of death was to inflict eternal suffering on the endless procession of its victims.

But it wasn't merely rats suffering the big-picture rationale. The U.S. Fish and Wildlife Service's ongoing eradication of arctic foxes from the Aleutian Islands, widely hailed as one of the paragons of island restoration, had been inflicting pain on both aggressors and victims. Ed Bailey, who had led the campaign for years, had often found himself in the predicament of a wildlife admirer obliged not only to kill but to work beside those who liked nothing better.

"I didn't relish eradicating foxes," said Bailey. "I knew it had to be done. But some of the people who went along—some of the Animal Damage Control people—they just got a big bang out of blowing away foxes. It was kind of disturbing to me. Some of the people in the fox camps just loved shooting things."

Callousness notwithstanding, the deaths by bullet were often merciful in their relative quickness. Those by leghold trap were often not. It was a dirty little secret among the fox-killing crews that the ethical standard of checking traps early and often did not apply in the Aleutian frontier. The Aleutian fox campaign was a bare-bones business of lowly paid trappers living in tents on lonely wind-battered islands on the edge of nowhere. The idea of checking every trap every day was understandably scoffed at by those facing daily marathons of boulder beaches and mountains of waist-high tundra. Bailey commonly came upon foxes dead in the traps, with untold hours and days of suffering exposed to the Aleutian elements. Foxes that didn't starve or succumb to exposure were sometimes cannibalized by their neighbors. Litters of pups whose nursing mothers didn't return home died a slow death in the den. "Leghold traps are cruel," said Bailey, "no question about it. They can say they're checking their traps ten times a day, but there's still a lot of pain involved."

Bailey in his tenure would come to whiff the scent of indifference from both sides of the ethical divide. In the 1990s the Aleutian fox project had a run-in with animal rights advocates from the Sea Shepherd Society. The Shepherds, better known for more high-profile escapades ramming whaling vessels and confronting clubbers of baby seals, had caught wind of the feds' fox trapping. The killing was cruel, they charged. The solution they demanded was to live-trap all the foxes and move them. Bailey, pondering the practical absurdity of their demands, called their bluff. When he offered to take the protesters out to experience firsthand the breadth and brutality of the Aleutian environs, the Shepherds turned tail and abandoned the foxes to their fates.

Bailey's only sure sanctuary of conscience was to remind himself repeatedly of the bottom line. "It was a dirty job that had to be done," he said. "But I had to look at the bigger part of the mission, of how many seabirds were alive because of what we were doing."

As for the vanquishers of Anacapa's rats, Howald and company would be saddled with the same imperfect solution to an inescapable dilemma. "Death is never easy," said Howald. "But we don't take it lightly. Unfortunately, this is the tool of choice we have."

## RESTORING BALANCE

One month after the quarreling parties of Anacapa came to legal blows, the federal judge handed down the court's opinion: The eradicators had followed the rules; they were free to proceed.

But by now the scuffle had gathered onlookers. Newspapers from Los Angeles to San Francisco picked up on the row between the murrelets' and the rats' defenders. Rhetoric-laden editorials and caustic letters to the editor frayed nerves. Anacapa rangers took to arming themselves with flak vests, handguns, and batons. Howald found himself looking under his car at the end of the day.

On December 5, 2001, the eradicators' helicopter lifted off with an industrial grain hopper and began methodically spraying East Anacapa with brodifacoum-laced pellets. Within hours the deed was done. After a few days of mop-up the island conservationists gathered in their rustic field quarters and popped the champagne. Said Tershy, raising a glass, "I think we made conservation history today."

Over the following year the checkup crews would find no signs of rats remaining on East Anacapa. But neither would they find any signs of the Anacapa deer mice that had been left outside. About a hundred birds were found dead too, ten of them raptors.

But from the death eventually came rebirth. In the spring following the poisoning, the captive deer mice were released, to soon replenish their island to capacity. Anacapa's cohort of young side-blotched lizards and slender salamanders had prospered particularly well during their first winter without rats. And that summer, in a cave where no Xantus's murrelet had fledged for the last three years running, surveyors found a nest bearing a healthy egg—the surprisingly rapid onset of a long comeback.

The following year's clearing of Anacapa's middle and west islets amounted to anticlimax. Again the helicopter sprayed; again the editorialists and letter writers blared foul. Deer mice, raptors, songbirds, and rats all suffered, but only the rats to the very last. And soon enough Anacapa had begun to resemble a more pristine past. By 2003 the number of murrelets nesting there had nearly doubled.

"That's why we're here," said Howald. "That's really why we're here. We're not here to kill rats. We want to see the sea-bird numbers take off. We're here for the lizards. We're here for the mice. We're restoring balance."

## Chapter 9

# ESCALATION

THE ERADICATION OF black rats on East Anacapa in December 2001 had capped a pivotal year for those who would save the world, one island at a time. It was a year rocked by wild swings between fortune and despair, beginning with fresh hopes for the poster victim of the global invasion.

The kakapo had weathered a tumultuous decade in the 1990s, replete with surprise stoat intrusions and rat-plundered nests and interisland shuttling of emergency patients, ever fleeing the reaper. And yet the turn of the twenty-first century revealed kakapo numbers in a tenuous upswing, having somehow risen from a moribund population of fifty-one to a veritable crowd of sixty-two. The upward inching of the kakapo population was a small but vital vindication for what had become a Herculean feat of human intervention.

In 1995, six years after Don Merton's last Fiordland expedition had come back having found no sign of any living kakapo, six years after Merton had begun lobbying for a heavier hand in righting the kakapo's fall, a decree came down from Wellington in the form of a new national recovery plan. In a case of better late than never, the hands–off philosophy was officially tabled in favor of immediate emergency aid. The direness of the situation was no longer to be ignored or understated. The kakapo was a species numbering in the dozens, with a glacial rate of reproduction and a world of enemies in firm command of its homeland. It had become a fugitive from its own country, and even the artificial shelter of island life had grown inherently dangerous. The refugees from Stewart Island, their numbers now squeezing into the bottleneck, were beginning to exhibit the classic death-spiral symptoms of inbreeding and failing fertilities. And any hopes of infusing fresh blood from Fiordland were hanging on one last old lovable bird who, for all anyone knew, might no longer be up to the task. For all anyone knew, Richard Henry, the kakapo's knight of knights, was a hundred years old. Such was the state of the kakapo's future when its rescue turned serious.

Under the new plan, every egg, chick, and kakapo mother was to come under intensive care. Tethered to every kakapo nest was a tent fifty meters away, hiding two stewards and a TV monitor tuned 24-7 to the kakapo channel. By day the stewards would review time-lapse tapes of the mother with her young, noting every turn of the egg, every feeding of a chick. Every coming and going of kakapo tripped an infrared sensor and the ringing of a doorbell in the watchers' tent. By night, as mother kakapo foraged, the stewards stood guard. Any rat attempting a burglary would first traverse a minefield of snap traps surrounding the nest. Any such rat lucky enough to survive that gauntlet would ultimately be met by the spying eye of the security camera and sent scurrying with a remote-controlled explosive charge. Or should, say, the cool of the night threaten to chill an untended egg, the stewards would intervene with a heating pad. If the kakapos' wild foods ran short, the birds found handouts conveniently appearing in their path. Every kakapo that had managed to survive to the mid-1990s was to find life a precarious but pampered affair indeed.

Punctuating the marathon vigils were moments to recharge the conservationists' flagging spirits, none more electrifying than the announcement commencing the 1998 breeding season, that Richard Henry, the most valuable kakapo on the planet, had finally mated. He, the last of the kakapo's Fiordland bloodline, and Flossie, a refugee from Stewart Island, had found each other in their temporary sanctuary on Maud Island. By early February, Flossie was on nest, incubating the first three eggs from which, everyone hoped, would hatch the heirs to Richard Henry's Fiordland.

Video surveillance and rat alarms, extra food and heating cushions, and all the sundry services routinely afforded every kakapo nest only began to describe the attention heaped on this critical little clutch. Flossie had chosen to nest on what to Merton and crew seemed a perilously steep and precarious slope. Over the next few mornings she would return from foraging to find her nest insulated with a fresh new bed of wood chips and bolstered by a plywood retaining wall with a viewing portal. At times she would return early and supervise the construction. Merton and his midnight remodelers dug drainage ditches, installed a new doorway, and built a deck. When the due date arrived, the remodelers turned midwives, whisking two of the pipping chicks into incubators to assist their labors. The team's exhaustive doting paid off. By the second week of March television crews were sharing images with the world of three homely, naked little hopes for the future of the kakapo.

And so, in harrowing fits and starts, evolved the seat-of-the-pants science of kakapo husbandry. As for securing the kakapo a safer interim home, the strategy had come to narrow on one island called Codfish. Situated off the west coast of Stewart

Island, a 3,400-acre, jungle-green throwback to primeval New Zealand, Codfish had been gradually groomed as the principal sanctuary in the kakapo's island shell game. In 1998 the final stage of Codfish's restoration commenced when the young wildlife officer Pete McClelland led a crew to kill Codfish's rats.

McClelland's feat on Codfish would presage the Americans' juggling act on Anacapa, and then some. Before a pellet of bait hit the ground, his crews rounded up and moved every last Codfish kakapo, more than four hundred rare bats, and twenty-one individuals of Codfish's unique brand of fernbird. Lacking anywhere suitable to store the fernbirds, McClelland cleared rats from two other nearby islands—in the same day.

Codfish's renovation was soon to be put to test. Early in 2001 the blossoming of the rimu trees on Codfish was hinting of a bumper crop of kakapo food, and therefore of kakapo chicks to come. Risking all kakapo eggs in one basket, their keepers hurriedly shipped all but a few infertile individuals to Codfish, in anticipation of what was hoped to be the biggest breeding season in the bird's new era of intensive care.

## AUCKLAND SUMMIT

That February, at the far end of the country in Auckland, the disciples of island conservation came together for the first major reunion since the Wellington rat conference of 1976. In as much as Wellington had been likened to the Last Supper for its somber undercurrents of impending loss, the spirit of Auckland would more aptly recall the Resurrection. This was no longer a surrender to fate, nor so simply circumscribed by one country's desperate affairs. In Auckland, the International Conference on Eradication of Island Invasives had gathered the big guns from what had become a global offensive.

There among them were Don Merton and Brian Bell, thirty-five years after bearing witness to Big South Cape's unraveling by rats. Both had since come to find their services in rising international demand, each now reporting on his most recent work clearing islands of interlopers in the Indian Ocean. Merton for his part, while not nursing kakapos back from the dead, had led a last-minute rescue of the Seychelles magpie robin, directing offensives against its attackers while carrying the last survivors of the species to safety.

There again were Bruce Thomas and Rowley Taylor, the now-renowned battlers for Breaksea, reviewing forty years of island rat eradications. The young industry's growth curve described a rocket's ascent, from the three-acre nubbin of New Zealand's Maria Island in 1961 to the eight-thousand-acre monolith of Canada's Langara thirty-four years later, with far more ambitious missions already counting down.

Pete McClelland was there to report that he and his Kiwi crew were just months away from a campaign to dwarf any in eradication history, preparing to remove every rat on the forty-four square miles of Campbell Island. Campbell, a world-class sanctuary of albatross and penguins 450 sea miles south of New Zealand, measured eight times larger and untold degrees more forbidding than Langara.

From across the continents came reports of innovative weaponry, mounting boldness, and once-invincible invaders being put on the run. Among the leaders of the Australian contingent, David Algar brought news of breakthrough in the Montebello Islands, having dropped eleven hundred sausages of kangaroo meat laced with Compound 1080 in an aerial eradication of feral cats. The Aussies had moreover been busy amassing an empire of recaptured lands, clearing foxes, rats, mice, rabbits, and goats from forty-five islands off their western shore since the 1960s.

There was Bernie Tershy speaking for his Baja band of island conservationistas, closing in on their twenty-fourth eradication in northern Mexico, their tally of rescued species surpassing fifty. And there too, ascending to the podium, was the feral figure of Tershy's unlikely ambassador, Bill Wood.

The man who had feared nothing more than to see his trapper's secrets unveiled before a packed auditorium—except, perhaps, the prospect of speaking in public—was now, at Tershy's prodding, doing both. And apparently loving it. The soft-spoken Wood, with his homespun lore and humble mastery of his craft, picked up where he had left off in the fishing villages of Mexico, now charming this international gathering of Ph.D.s and professional wildlife managers. Wood was once again the unassuming sage of laconic wit, the Yoda of the cat wars. In the Q&A that followed, when asked where to best set a cat trap, Wood brought the house down with the wisest three words a master trapper ever divulged: "Beneath its foot." Afterward he found himself in the familiar situation, the magnetic center to a gathering circle of curious strangers and instant friends, fielding offers of home-cooked meals and places to stay.

It was a conference for both the graybeards and the young guns of island conservation, some coming face-to-face for the first time, some leaving the meeting with audacious new plans. Josh Donlan, the ponytailed, Birkenstocked, self-described hippy sort from Tershy's Santa Cruz team, and Karl Campbell, a hard-charging, crew-cut grad from Queensland rapidly gaining a reputation as the goat-killer of the Galápagos, had from their e-mail correspondence each imagined the other as a grizzled, battle-scarred veteran of the island wars. In Auckland they met to their surprise as fellow twentysomethings, and as kindred souls the two immediately began plotting world conquests.

Campbell had been developing a diabolical means of matching

wits with one of the toughest and craftiest of the island invaders, even then eating its way through the hallowed kingdom of the Galápagos. There had been many attempts around the world to eradicate island goats, many of them failing for one particular reason. The hunters, with their guns and dogs, in their opening salvos would invariably mow through the unwitting herds. But those few that would typically escape would seem to sprout the capes of supergoats. Time and again the hunters would be left chasing wary phantoms vanishing into the island's most rugged enclaves. And all it took was one gravid nanny to outlast the hunters' money or patience, and the island's goat invasion would mushroom all over again.

Campbell capitalized on one of the tenets of goat warfare, which stated that the surest way to lure a goat from hiding was with another goat. Judas goats, they'd come to be called. Collared with radio transmitters, Judas goats were sent out to seek and lure herdmates while silently signaling their whereabouts to the hunters. The Judas goat had revolutionized the hunt for those last island holdouts and renegades; Campbell took the weapon of seduction one step further. He began implanting nannies with extra doses of estrus hormones, creating in essence a supersexed she-goat that had her suitors running from cover. Campbell's hoofed femme fatale drew irresistible parallels to a famous World War I spy, a female exotic dancer doubling as an agent for Germany and eventually executed by firing squad. Mata Hari, they named Campbell's monster.

Within four years of meeting, Campbell the hit man and Donlan the strategist would lead two hundred ground hunters, forty trained dogs, six hundred Judas goats, and an ace squad of helicopter pilots in a campaign covering a combined landmass the size of Rhode Island. With guns ablaze, firing over half a million rounds of ammunition and killing 150 invaders per hour, the team and their Ecuadoran counterparts would remove every last one of 160,000 goats from the Galápagos islands of Santiago and Isabela, the two biggest goat eradications in history.

## SHOULD WE BE WORRIED?

Away from all the plenary speeches and major announcements of record-setting restorations, in a quieter corner of the convention hall, stood a large man beside a poster summarizing his research. Art Sowls was a seabird biologist with the U.S. Fish and Wildlife Service, his territory covering the Alaska Maritime National Wildlife Refuge. Nine months earlier Sowls and a fellow seabird biologist, Mark Rauzon, had journeyed to the outer reaches of the Aleutian chain, to the legendary black tongue of lava at Kiska's Sirius Point and the greatest single colony of auklets on the planet. What they had found there amply confirmed fears that by then had been building for twelve years. Sowls's poster was titled "Can Kiska's Auklets Survive the Rat Menace?"

The first hints of trouble at Kiska had surfaced in 1988, with what had begun as a routine survey of the auklets of Sirius Point. One of a trio of biologists, Hector Douglas, was off hiking when he came upon the track of what could only be an arctic fox. It was a fox that should not have been. Hired hunters had supposedly finished clearing the last of them from Kiska two years before.

Douglas returned to the auklet colony to mention his discovery to workmate Dave Backstrom, who by then was puzzling over a strange sighting of his own. Backstrom had been watching as a pocket of melting snow in his study plot began to unveil something odd. He brushed aside the snow to find a bundle of molding auklets, each hardly ruffled but for a hole in the head where the brain had been. Backstrom and Douglas's disparate findings suddenly clicked. The two radioed notice of the renegade fox to the refuge managers, who dispatched a trapper, who dispatched the fox. And that appeared to settle things: The cache of auklets had been the work of Kiska's fugitive fox, thought the experts. The fox was now gone; crisis averted. And for a while, nobody thought anything more about it.

In the years thereafter, biologists periodically visiting Kiska would continue to find fresh stashes of dead auklets, but no more signs of foxes. Sowls and Rauzon were to follow in 2000, now with curiosities backed by mounting concerns. They repeated the steps of their predecessors, slipping ashore between the poundings of treacherous breakers, climbing onto the devil's playground of volcanic boulders, wandering for a week about the rooftop of the ultimate auklet metropolis. There they came upon the same macabre scenes, of auklet corpses stacked like cordwood, their brains and eyeballs eaten out. They found too the remains of eaten eggs, chicks, and what had been parents tending their nests. The biologists set out quail eggs and returned the next morning to find a third of them destroyed by rats.

Sowls and his fellow caretakers of the Aleutians had wanted to believe that the auklets in their multitudes would simply swamp whatever carnage the rats of Kiska could muster. But the fresh bodies at Sirius Point now had them wondering. Was the epic flock too massive for even the indomitable rat to chew through? Or were these the mass graves of a great spectacle on its slide into oblivion? Presenting these scenarios to the international gathering of island experts in Auckland, Sowls in 2001 stood before his fellow conferees and asked, "Should we be worried?"

If Sowls had come looking for reassurances, he had come to the worst possible place. He was surrounded by Kiwis who had spent careers rebuilding their island nation from the rubble of the rat bomb. The bird of Sowl's concern was, after all, a chunky, defenseless morsel whose inherent preference for nesting in crevice-riddled rock could not better suit a rat. "Mate, you are out of your mind not to worry," Sowls heard time and again. "It might not happen overnight, it might take fifty years or more, but you'd better believe that as long as rats live on Kiska, your auklets are in danger."

Four months later, scientists embarked on the research vessel *M/V Tiĝlax̂* of the Alaska Maritime National Wildlife Refuge, heading west across the Aleutians to begin a thorough reconnaissance, deciphering who between the auklets and the rats was winning the mortal contest on Kiska. And wondering what, if anything, might be done about it.

*Chapter 10*

# SIRIUS POINT

IAN JONES WAS a forty-one-year-old professor of biology from Memorial University, in Newfoundland, Canada; an eighteen-year veteran of seabird study in the Bering Sea; and considered by most to be the world's leading authority on the least auklet. His foremost subject was an incredibly common bird remaining largely unknown, for good reason.

As an individual, the least auklet was easy to overlook. At a mere three ounces it was among the tiniest of seabirds, and the reigning runt of a family of web-footed, flipper-winged, tuxedo-plumaged parallels to the penguin. On its nesting grounds, however, the least auklet grew overwhelmingly conspicuous. "They swarm in millions about the rocky beaches of the Pribilof Islands, outnumbering any other species in the Bering Sea," wrote the early-twentieth-century ornithologist Arthur Cleveland Bent. "It is difficult for one who has not seen them to appreciate their abundance and one is not likely to overestimate their numbers." When it came to conducting one of the most basic field measurements of the wildlife biologist, the best one could hope for with regards to the least auklet was to sample small patches, extrapolate wildly, and throw up one's hands.

So it went for much of the least auklet's life history. The auklet held the maddening distinction of being the most numerous seabird in the northern hemisphere and perhaps the least understood. In summer it appeared in its staggering flocks upon the breeding rocks of several dozen islands. For the rest of the year the auklets in their uncountable millions all but disappeared.

They spent the long gray months somewhere in the vastness of the North Pacific, diving, hunting, sleeping, and enduring storm and swell. With the final days of April the auklets would amass from their secret corners of the sea and flock shoreward to breed in their legendary congregations. The birds would alight on the rocks of talus fields and lava flows to bob and weave, strut and court, their excited, buzzing chatter coalescing to a roar audible more than half a mile away. And just as suddenly they would vanish, ducking below to nests where human hands and eyes could seldom reach or see, to lay a single egg that they would take turns incubating for a month to see it hatch. A person standing atop a city of millions in the middle of the day might come away believing the place deserted. With morning the birds would fly out to feed, disappearing again underwater, again to where few scientists would ever have the good fortune of observing them in a most fundamental act of feeding. Dissected auklet stomachs and purged throat pouches revealed that the little birds chased swarms of tiny antennaed creatures called copepods, and at a frantic pace, consuming upward of 90 percent of their body weight each day. Come August, with chicks fledged and gone to sea (a harrowing affair, involving a shaky first dash for the water, followed by an independent life on the waters of one of the harshest seas in the world), the great colony would break up and head back out to god-knew-where. Jones, who went on to write the definitive monograph on the least auklet's life history, could muster only three sentences on the bird's winter range and migration, summing up the state of the science as "poorly documented" and "little known."

The elusive least auklet was nonetheless easy prey in the breeding colonies. Auklets emerging from their crevices or hesitating on a rock before takeoff were regularly swallowed by gulls or picked off by snowy owls. Auklets taking flight suffered the stoops of bald eagles and peregrine falcons from on high. Those swimming offshore were sometimes swallowed by fish or seized by the relentless gulls. Humans too hunted the auklets. Aleut people in the Pribilofs had been known to bag hundreds by the hour, by expertly waving a net as the flocks zipped past.

In his 1993 monograph Jones did broach the subject of foreign predators. The era of Aleutian fox farming had graphically exacted its toll on the littlest of auks, perhaps eliminating colonies from the islands of Uliaga and Kagamil. The impact of the rats, though, was only lately coming to light.

Following the near-eradication of Kiska's foxes in 1986, visiting biologists would note rat tracks and droppings proliferating on Kiska's beaches. Whether the rats in their freedom were multiplying or simply venturing more boldly with the coast finally clear—perhaps it was simply a lack of somebody there to notice—one could only guess. Whatever the case, there was something new afoot in the foxes' absence. When the first bundle of birds was found lying so neatly stacked during the Sirius Point auklet expedition of 1988, it was still possible to blame the last renegade fox. With the last fox's subsequent killing, the list of suspects narrowed considerably. When in 1992 the refuge's chief biologist, Vernon Byrd, and his fellow seabird specialist Art Sowls came upon the same odd spectacle—this time twenty-eight auklets laid side by side—there was only one viable predator left to blame.

"Accidental introduction of rats to remote islands (due to shipwrecks or at harbors) may be the single most serious threat to Aleutian and Pribilof breeding populations," wrote Jones in 1993, conveying Byrd and Sowls's concerns. Nine years later he would see for himself.

In the summer of 2001, Jones began what would become a perennial study of the auklets and rats of Sirius Point. He followed the familiar migration route of a small flock of scientists conducting research each summer across the outer Aleutians, flown by a DC 737 from Anchorage twelve hundred miles west into the Bering Sea to the old navy airstrip on Adak Island, there boarding the *M/V Tiĝlax̂*, the 120-foot workhorse research vessel of the U.S. Fish and Wildlife Service. Twenty-four hours west of Adak's Sweeper Cove, the *Tiĝlax̂* came to anchor off the famous lava flow marking the northernmost shore of Kiska. Ferried from there in an 18-foot inflatable, threading warily between breakers through a notch of rock, Jones jumped to shore at the base of the Sirius Point auklet colony.

His home for the next three months would be a campsite shoehorned into the rocks, his communication consisting of two radio conversations each day with refuge personnel on Adak. His accoutrements featured a mountain tent for sleeping and a twelve-by-eighteen-foot nylon shelter for eating, studying, and otherwise hunkering down for however long Kiska's weather so dictated. The shelter was strutted with steel, bolted to a wooden foundation, and anchored six ways 'til Sunday against the weekly Aleutian apocalypse. It was custom-built to withstand even the one-hundred-mile-per-hour williwaws that would come roaring unannounced from the heights of the Kiska volcano. For protection from the pounding northerly gales, the campsite had been tucked behind a fifty-foot wall of rock fronting the sea, a rock that stopped most of the waves.

In his first summer Jones began laying the groundwork, hiking about the slippery jumble of boulders (a feat once compared to walking on giant marbles), marking out his study plots, erecting his blinds, catching auklets on a carpet of nooses, cinching colored bands around their legs, and watching.

Every day, twice a day—when the gales and williwaws permitted—Jones headed out to the blinds. And from nine in the morning to two in the afternoon and from ten at night to half past midnight, Jones sat peering out upon the rocks, charting the comings and goings of the resident parents and watching for the defining finale of lone chicks clumsily departing for the sea. As a gauge of Kiska's relative normalcy, Jones had students stationed at auklet colonies on two other Aleutian islands, Buldir and Kasatochi, watching likewise. If something bad were going on at Kiska, it would likely show by contrast with the ratless colonies on Kasatochi or Buldir.

Hardly a week into searching the lava dome of Sirius Point, Jones came upon a cache of thirty-eight least auklets, their bodies apparently moldering from neglect. Eleven days later he found another little massacre of at least four, barely visible from the depths of a crevice. They too were decomposing, left to rot by a rat that had apparently gone on to other things. That year, barely more than one out of ten auklet nests at Sirius Point hatched a chick. By contrast, auklet chicks on Buldir and Kasatochi were flying away at three to five times the frequency. The breeding season of 2001 ended with the worst production of least auklets Jones had ever witnessed anywhere.

The following year Jones was joined by his grad student Heather Major, and the two repeated the experiment. The summer started off with an unfortunate bang. On May 26, 2002, the two came upon a stash of 122 least auklets, together with the corpses of seven fork-tailed storm petrels, another little burrowing seabird sharing the lava flow. In the burrow lay a female rat with nine pups. Through June the bodies continued to pile up. On the 29th, Major and Jones found a single cache of decomposing bodies numbering 148 least auklets.

Wandering about on the lava dome, Major and Jones came upon thousands more carcasses and half-eaten eggs and what Jones would describe as "windrows of skulls," each with a hole through which the brain had been extracted. That year beat the previous year's record for the worst auklet production in history. "It is an enormous disaster," Jones reported. "The number of seabirds that are being killed by rats each year are more than what were killed by the Exxon Valdez oil spill."

The colony was under siege. And the plague had developed a certain disturbing progression. All the hoarded auklets were identified as adults killed early in the breeding season. As the season matured, the hoarding stopped, the caches giving way to the remains of chicks strewn about the colony. Jones and Major found evidence of auklet parents ambushed at the entrance of their den, their food pouch spilled on the rock, their chick eaten in turn.

The developing picture was that of a colony of half-starved rats barely hanging on through the cruel Kiska winter, then finding with the auklets' spring arrival their manna from the heavens. Obeying instincts forged by eons of feast or famine, the rats immediately set upon the windfall of auklets as if it were the last meat they would see for months. Once sated on eyeballs and brains—the richest organs in the package—the rats switched gears from famine mode to reproduction mode. And soon the rocks of Sirius Point were visibly crawling with rat pups.

Although the surplus slaughters lasted but a short while, their timing and intensity were especially crushing for the auklets. The birds most vulnerable were the first arrivals, the dominant, most experienced breeders, the ones most capable of seeing a chick through to fledging. The starving rats in their desperate hoarding were skimming the reproductive cream of the auklet crop.

Major did the math and came up with a sobering model of probability. At the going rate the rats could nearly obliterate the incomparable auklet colony of Kiska within thirty years. Major and Jones were later joined by refuge biologists Vernon Byrd and Jeff Williams in an official message to the scientific community, published in the journal *Auk*. "The presence of introduced rats at Kiska is of great concern," they wrote, "and we recommend their ultimate eradication."

# Chapter 11

# RAT ISLAND

As the fears for Kiska mounted with the carnage, those tending the Alaska Maritime National Wildlife Refuge were already busy considering the deeper ramifications. In their jurisdiction were at least a dozen islands with rats chewing holes in the ecosystems they were charged with defending. The concerns went far beyond seabirds. It was about the missing songbirds and Kamchatka lilies, the shorebirds and the dune grass and the intertidal communities languishing in the post-rat age of the Aleutians.

And now Kiska, among the finest jewels of the chain, was being robbed beneath their noses. Art Sowls, in his years defending the harbors of the great Pribilof rookeries against incoming rats, had come to assume that prevention was the islands' only salvation. He and his boss, Vernon Byrd, had repeatedly characterized the invasion of a single rat as a fate worse than any oil spill.

Prevention on Kiska was obviously no longer an option. Neither was neglect. Sowls and Byrd had seen for themselves the corpses. They had heard their worst-case scenarios emphatically confirmed by the experts in Auckland, and now by the world's leading expert on auklets, staring directly into the rats' lairs at Sirius Point. The question was no longer whether Kiska needed saving, but how.

In the forty years since Bob Jones had committed himself to taking back the Aleutians, he and his followers had cleared foreign foxes from forty islands. They'd seen seabirds piling back into the predator-free vacuums; they'd celebrated the Aleutian cackling goose rebounding from assumed extinction to some thirty thousand birds, flying itself off the U.S. register of endangered species. But here was a scarier foe, a more suspicious creature than the curious fox (which had a reputation for helping itself into traps and walking up to men aiming rifles). The rat presented logistical problems far beyond the purview of a few underpaid hunters laying lines of leghold traps by hand.

Up until 2001 the stewards of the Aleutian refuge had considered the invasion of rats as the sentence of an incurable cancer. That year, even as the auklet bodies were stacking like cordwood, Kiska's terminal prognosis took a heartening turn.

## McClelland's Campaign

In July 2001 a team of New Zealanders under the direction of the wildlife officer Pete McClelland reported that they had just poisoned what they hoped was every last rat on an island larger by far than any ever attempted. Campbell Island, a forty-four-square-mile, starkly beautiful, foul-weather paradise in the Southern Ocean of the subantarctic, had become the giant new standard in the campaign for island conservation.

Unwelcoming by all outward appearances, Campbell had nonetheless come to host a massive gathering of wildlife. Great colonies of penguins, albatross, petrels, shearwaters, prions, and cormorants and immense rookeries of seals and sea lions gathered in Campbell's paradise of seclusion. And all had eventually come to suffer the familiar plague of discovery. By the late 1700s, Campbell Island's seals had brought the inevitable shiploads of seal hunters, and with them their rats. The sealers bludgeoned their way through the herds; the brown rats chewed their way through the flocks.

McClelland, charged with saving the endangered, had two monsters to deal with. One was the rat; the other was Campbell Island. Campbell had long been deemed undoable for good reason. Bounded by thousand-foot cliffs, it sat far alone in the Southern Ocean—just getting there meant crossing 440 miles of heaving seas through the storm-lashed latitudes so reverently named the Roaring Forties and the Furious Fifties. McClelland spent the better part of five years planning and organizing for what would closely approximate a military invasion. It would be a blitzkrieg without compromise. If one pregnant rat were left in one little cranny of Campbell's rugged enormity, future eradications, careers, and two million dollars of taxpayers' money would be lost. Failure was not to be discussed. Any naysayer among McClelland's crew was immediately surrounded and shouted down. It was better to lower the head and trudge forward than to look up and contemplate the impossibility of the looming task. Over the final six months of preparation, McClelland took to sleeping with a notebook next to his pillow, jotting down scraps of thought and reminders as they woke him through the night.

On June 26, 2001, the beginning of the New Zealand winter, five helicopters and two ships loaded with 210 drums of helicopter fuel and 132 tons of rat bait left Invercargill, at the southern terminus of New Zealand, heading for the Southern Ocean. By the second week of July, McClelland and company had dropped their first load of brodifacoum on Campbell. Ten days and two hundred thousand dead and dying rats later, they dropped their last. In the following years the crews returned to find no evidence of survivors.

Campbell's eradication raised the bar by frightening degrees. The island exceeded the area of former record-holder, Langara, by eight times. McClelland, who took three years before returning to Campbell, had trouble believing what he'd done. "I looked at those cliffs and the size of that island and I said, 'No. There's no way we did that. It was too big, too rugged to get rats off that island.'"

Campbell Island came as both fresh hope and forewarning for those now contemplating Kiska. In many ways Kiska was Campbell's big brother of the north. Both lay at the high latitudes nearing fifty-two degrees. Both were far adrift in forbidding seas, pummeled daily by winds gusting to fifty miles per hour, and stung by cold, incessant rains. There was good reason that neither was inhabited or blithely happened by. Stormy spells could last weeks, when landing a boat or flying a helicopter would threaten lives. Yet somehow, on Campbell, McClelland and the can-do Kiwis had managed to breach the impenetrable fortress. To the minds of their North American allies, if Campbell, why not Kiska?

## TARGET PRACTICE

Conversations began. Sowls and Byrd of the Aleutians, to McClelland of the New Zealand territories, to Gregg Howald of Island Conservation—together they started pondering first steps toward tackling Kiska, the very first of which was to go nowhere near it.

Attacking Kiska outright was a fool's game. Kiska was twice the size of Campbell, and of any rat island ever attempted. It was one hundred square miles of fog-shrouded rock and tundra, fifteen hundred air miles from Anchorage, plus a twenty-four-hour cruise through cranky seas from the nearest working harbor. To reach Kiska alive was only half the ordeal. Inescapably there was that volcano to deal with, a snow-covered monolith whose upper reaches remained hidden for all but a few days by the Aleutians' notorious gray curtain of cloud. And at the base of it lay ground zero, Sirius Point, that all-but-impenetrable maze of Volkswagen-size lava boulders, tantamount to the world's largest bomb shelter for rats. Kiska was the worst of places to pick the first rat fight in the Aleutians.

The sane approach would start smaller, someplace challenging but manageable, someplace not so little as to be yawned at, not big enough to risk humiliating defeat. As McClelland would warn, "If you don't succeed, you don't get the chance to do another."

Among the chain of Aleutian candidates, one particular island all but nominated itself by name. It had been infested since 1778, when a Japanese fishing vessel had run aground on its reef and spilled a few rats ashore. The native Aleut people had earlier named the island Hawadax, meaning "entry" or "welcome." When the Russian explorer Fyodor Petrovich Litke came by in 1827, he renamed it for what had by then apparently become its most striking feature. Litke named it Rat Island.

Rat Island was but a tenth of Kiska's size, but at seven thousand acres it would rank as the world's third largest rat eradication ever attempted. Rat Island was classically Aleutian: windy and cold, thirteen hundred miles west of Anchorage, and a twenty-four-hour boat ride from the closest port, at Adak.

It was not a venture for the fainthearted, but victory there promised big returns. The Aleut people had left indications of what could be expected. Long ago they'd piled their trash heaps high with the bones of seabirds. Their ancient middens revealed a former avifauna of impressive diversity, including puffins, petrels, murrelets, gulls, and cormorants. All were species that would presumably come spilling back once Rat Island was rescued.

By all accounts, the island would provide a challenging warmup for the title bout with Kiska. "It was a mama bear, papa bear kind of thing," said refuge biologist Jeff Williams. "Not too big, not too little. We knew from our earliest stage of thought that if we were going to do an island, it would have to be Rat Island."

In the summer of 2001, Williams, his refuge colleague Sowls, and Howald visited Rat Island for an early reconnaissance to see for themselves what they might be getting themselves into. How big and steep and manageable—or not—was the terrain? Where might the ground crews camp? Where on the tundra slopes would the helicopters land and load? How bad, really, were the rats?

All three were well traveled and intimately familiar with some of the great seabird colonies of the Pacific—with the quintessential cacophony and commotion of birds, the acrid essence of guano. When they landed their Zodiac inflatable on the beach, the overwhelming essence of Rat Island was that of silence and sterility.

All knew from looking at the empty cliffs that this was a place that should have been far busier with the air traffic of tufted puffins and ancient murrelets, the headlands pocked with the burrows of whiskered auklets and storm petrels. "Rat Island was almost a dead zone," said Howald, "except for the rats."

That night, as the three pitched their tents, Howald, knowing better from his experience, steered clear of what was sure to be a riot come nightfall. He shouldered his gear and headed up the hill for a sheltered spot in the tundra. Sowls would later wish he'd followed. Sowls in a lazy moment decided to forgo the hike with Howald and brave the beach. As a precaution he surrounded his tent with snap traps. He zippered down the door, crawled into his sleeping bag, and, almost before his head hit the pillow, jumped to the *bang* of a trap going off. The next morning Sowls climbed outside to find all six traps clutching dead victims. One of the rats had been stripped to the bone, cannibalized by its cohorts.

Sowls and company returned from their little adventure, and the talks about taking the rat out of Rat Island got serious. They had seen up close an impoverished ecosystem needing repair and the stepping-stone to the biggest endangered prize in the Aleutian refuge, Kiska. It turned out that their seat-of-the-pants impressions from a few days' visit, of the destructive power of a novel predator unleashed, were even then being backed by a spate of new science, some of it conducted in familiar surroundings.

## THE RAT KING

That summer another team of biologists descended on the Aleutians, with the idea of testing, in particular, a well-informed hunch about foxes and seabirds and, more broadly, an emerging curiosity of science known as the ecological cascade. The cascade was a sequence of life and death triggered by disruptions at the top of the food chain, and it appeared to be playing out in grand fashion across the Aleutians. Veterans of the islands had come to recognize a certain qualitative difference between those still harboring introduced foxes and those without. It wasn't just the missing seabirds; it was something about the landscape. The fox-infested tundra seemed browner, more barren. It seemed to lack that certain profusion of Kamchatka lilies and tall green swards of cow parsnip and seacoast angelica. More foxes to fewer seabirds to duller landscapes, went the hypothetical cascade. And the converse.

To test their hunch, the biologists hopscotched across the Aleutians, sampling islands with foxes and without. If they could land a skiff without killing themselves, they sampled it.

On uninvaded islands like Buldir and Chagulak, with seabirds amassing by the millions, the avian populations outnumbered those on fox islands by orders of magnitude. Where the foxes had invaded, the soils supported a crusty cast of mosses and lichens and low-lying crowberry shrubs, an impoverished, poorcousin rendition of the Aleutian maritime tundra. Yet where the seabirds still reigned, so rained their guano, rich in nitrogen and phosphorous, the food of plants. The vital nutrients measured sixty-three times richer on islands without foxes. The fertilized soils grew tall grasslands, blooming with those Kamchatka lilies and towering green herbs. The seabirds were no bit players in the Aleutian ecological theater. They were chief gardeners of the maritime tundra, transformers of landscapes, and conspicuous in their absence.

Which raised again the question of rats. They were as likely as foxes to be cleaning out Aleutian seabird colonies, to be starving the tundra. There were hints that they were cutting swaths in the marine communities as well. One hypothesis held that rats, by driving away gulls and shorebirds from the Aleutian beaches and tide pools, were allowing the birds' prey to flourish to destructive densities. Freed from pecking beaks, snails and limpets grazed unhindered through the coastal kelps and seaweeds, gutting a critical organ of the intertidal ecosystem.

It seemed that wherever inquiring minds looked, the rat's reputation as ecological kingpin was gaining scientific weight.

One such place was the Hawaiian island of Oahu, in an ecological enigma called the Ewa Plain. The plain was a grassy expanse once covered in a rich forest of palms. Botanists for years had blamed the missing forest on the fires of the colonizing Polynesians, if not on the axes of Captain Cook and his successors. By the late 1990s, archaeologist Stephen Athens had good reason to believe that those supposed fires and axes had in fact featured four legs and a naked tail.

Athens in his excavations found that the palm forests had begun collapsing in about A.D. 1020, some four hundred years *before* the first Hawaiians had come to settle there. Athens postulated that the rats, arriving in the canoes of the first Hawaiians, and finding themselves loosed in a predator-free land of plenty, had swelled to monstrous masses and rolled like a living tsunami across Oahu, hitting the Ewa palm forest long in advance of their people. Once arrived, the swarms lived large on palm nuts. No more palm nuts, no more palms. Without spark or blade, the forest of the Ewa Plain had toppled of a rat-borne decay.

Hawaii was neither the last nor the least of the Pacific islands to have its popular rendition of history shredded by rats. Something similar seems to have happened on a sixty-four-square-mile protrusion of treeless terrain two thousand miles west of Chile named Rapa Nui, more popularly known as Easter Island. Originally rising to fame on the riddle of its stone-faced monuments—thirty feet tall and eighty tons heavy, staring blankly out to sea from the middle of nowhere—the island's reputation had more lately come to center on the culture that had placed them there. Easter Island had become a parable of humanity's ecocidal tendencies, largely owing to the vivid interpretations of the scientist and bestselling author of *Collapse* Jared Diamond: "In just a few centuries, the people of Easter Island wiped out their forest, drove their plants and animals to extinction, and saw their complex society spiral into chaos and cannibalism. Are we about to follow their lead?"

Diamond's sobering summary involved Easter Island's Polynesian settlers cutting forest faster than it grew. As the trees ran short, so did the fuel of their fires and the fabric of canoes and houses and livelihoods. Life got hard, war broke out, society collapsed. "As we try to imagine the decline of Easter's civilization," wrote Diamond, "we ask ourselves, 'Why didn't they look around, realize what they were doing, and stop before it was too late? What were they thinking when they cut down the last tree?'"

In 2007 the archaeologist Terry Hunt, from the University of Hawaii at Manoa, and his colleague Carl Lipo, from California State University, Long Beach, answered the question for the missing Polynesians. Maybe, argued Hunt and Lipo, those felling the last tree did so with neither ax nor aforethought. Maybe Easter Island had fallen not to short-sighted humans, but to a feeding frenzy of Pacific rats.

Hunt and Lipo's scenario of Easter Island's demise, like Athens's of the Ewa Plain's, began with the arrival of the Polynesian voyagers around A.D. 1200, accompanied, of course, by their rats. The rats proliferated in a predator-free paradise, finding millions of palm trees, each producing an annual crop exceeding 250 pounds of nuts. A single mated pair of rats, on an island with limitless food and no predators, could double their numbers every forty-seven days, becoming seventeen million in three years. Almost all the plants extinguished from Easter Island would prove to be rat favorites. One shrub that did survive featured a seed that seemed to germinate better for having been chewed.

In 2006, Hunt and Lipo published their solution to Diamond's riddle by way of the journal *Science*. "The ecological catastrophe of Rapa Nui had a complex history that cannot be reduced to psychological speculations about the motivations of people who cut down the last tree. Indeed, the 'last tree' may simply have died, and rats may have simply eaten the last seeds. What were the rats thinking?"

Those still harboring doubts about the rats' omnipotence were to be directed to another recent paper, this one beginning with the far broader question: "Have the Harmful Effects of Introduced Rats on Islands Been Exaggerated?" In their answer, David Towns, from the New Zealand Department of Conservation, and two colleagues produced a withering compilation of extinctions and decimations, of birds, bats, bandicoots (a rabbitlike cousin to the kangaroo), worms, insects, spiders, crabs, frogs, snakes, lizards, shrews, and shrubs, all with histories of disappearances linked to the arrivals of brown rats, black rats, and Pacific rats. The trails of carnage led from one-acre islands in the French West Indies to the 607 square miles of Hawaii's Oahu, across the Caribbean, Mediterranean, and Tasman seas, across the Indian, Atlantic, and Pacific oceans. Any more questions? In yet another synthesis of damages, researchers from the University of Hawaii at Manoa seconded the Towns report, crediting the big three species of rats with having driven or assisted 103 species of birds, reptiles, amphibians, and mammals to extinction.

Even the humble house mouse—of the species better known as invader of pantries and kitchen cabinets—had recently unveiled an alter ego as voracious slayer of island giants. Far off the southernmost coast of South Africa on Gough Island, home of the last breeding colony of the Tristan albatross, video cameras trained on their nests captured what had to be seen to be believed. Gangs of mice were rushing from out of the dark to attack birds three hundred times their size. The mice were chewing holes in the rumps of seventeen-pound albatross chicks as they sat, eating the living birds from the inside out.

And no modern chronicle of rodent overachievement would be complete without the saga of Razza the rat. Razza was a wild brown rat captured, named, and released in November 2004 under intensive watch by a team of biologists led by the University of Auckland's James Russell. Attempting to measure the difficulties of capturing or even detecting a single invading rat, Russell's team radio-collared Razza, turned him loose on a twenty-three-acre island in northern New Zealand's Hauraki Gulf, then set about tracking him down.

Razza had other ideas. For weeks he repeatedly refused all invitations to plastic tunnels and snap traps and turned down all sensory enticements from fish oil to chocolate. Then his radio signal vanished. To the embarrassment of his watchmen, Razza had escaped.

Following a tip from villagers on a neighboring island, Russell and crew went looking. And there, to their astonishment, they found a rat scat, identified by DNA fingerprinting as Razza's. The roving rat had swum a quarter mile of open water.

Again the scientists gave chase, this time siccing trained dogs on Razza's scent. And for another six weeks, the rat continued to run rings around them all, until finally—four and a half months into the chase—with a momentary lapse of caution and a meat-baited trap—the odyssey of Razza came to its dead end.

Russell and colleagues concluded their part of the scientific adventure with a newfound sense of respect and a healthy serving of understatement: "Our findings confirm that eliminating a single invading rat is disproportionately difficult."

## Quietly Conserving Nature

And so the rat's reputation as supernatural escape artist and ecological wrecking ball expanded in tandem with the island conservationists' escalating offensive. The state of eradication art
had taken great leaps in the forty years since the birdman Don
Merton and the schoolteacher Alistair McDonald had almost
haphazardly cleared the nubbin of Maria Island with a few
handfuls of warfarin. There were now helicopters navigating by
satellite, air-dropping poison by the tens of tons, striking with
precision measured to the meter. The eradicators' ranks included
hormonally engineered pigs and Mata Hari goats diabolically
luring herdmates to slaughter. Islands approaching the size of
small states and archipelagos spanning the seas were being rear-
ranged and restored to more pristine pasts under strategic assault
and rescue.

The fight had spread far beyond the pioneering Kiwis and
Aussies in their embattled island nations. The managers of the
Alaska Maritime National Wildlife Refuge had cleared upward
of forty islands of foreign foxes, resetting the stage on the most
prolific seabird chain of the northern hemisphere. The crews of
Island Conservation and their Mexican allies had conquered
invaders across two dozen islands of Baja, with their sights on
far broader horizons. They were squaring off against mink and
raccoons in the Scott Islands of British Columbia, mice in the
Farallons of California, macaque monkeys in Puerto Rico, and
goats, cats, and rats in the Galápagos.

And now all attentions had come to focus on Rat Island, and
the pending attempt at the third-largest rat eradication ever, in
a devil of a destination, with a price tag running into the millions. Its planners readily assembled the major components. In
brodifacoum they had a proven poison, and in Pete McClelland

they had ready advice from the world's record holder at rat eradication. With the U.S. Fish and Wildlife Service they had the keys to the refuge and the comradeship of the most knowledgeable sailors and pilots and naturalists of the Aleutians. And soon they had the millions too.

Rat Island's eradication went from dream on a drawing board to economically viable work plan when the Nature Conservancy joined the campaign. The Nature Conservancy—owner and steward of the largest private network of nature sanctuaries in the world (a wild kingdom spanning more than a hundred million acres of premier ecological real estate in some twenty countries) and perennially ranked among the top twenty philanthropic organizations in the world—knew a thing or two about orchestrating grand conservation ventures. Moreover, the conservancy as of 2004 had just co-authored a plan with the World Wildlife Fund with an ambition no less audacious than to save the Bering Sea. Tops among the conservation targets were the Aleutian Islands and their unprecedented gathering of seabirds. And chief among the solvable threats to those priorities were predatory invaders. Rat Island was tailor-made for the conservancy's first big splash in a Bering Sea campaign.

The Nature Conservancy came with cash and cachet. It also came with fresh scars from its forays into the emerging practice of conservation by eradication. In as much as the bulldozer had once served as symbolic antagonist to the conservancy's mission, choking weeds and animal pests had since invaded their way to the top of its most-wanted list. Proudly touting the motto Quietly Conserving Nature, the organization had over the years found it ever harder to protect its investments without unwanted noise, in such forms as gunfire and public outcry.

The conservancy's first major foray into the killing arena began in 1978, when it purchased all but 10 percent of Santa

Cruz Island, off the coast of Southern California (the rest was eventually incorporated into Channel Islands National Park). The property conveyed a host of rare and indigenous plants and animals, as well as a legacy of livestock threatening to displace them. Cattle, sheep, and pigs had trampled and chewed the island, sending hillsides sloughing into the sea and the island's native life-forms running for cover. Taking inventory of the endangered, conservancy steward Will Murray one day found himself hanging from a rope on the face of a Santa Cruz sea cliff, counting the last three survivors of a wildflower species driven to the edge by marauding sheep. Thereafter Murray's duties as Santa Cruz steward included shooting sheep. The conservancy's new tack of quietly killing for nature went public rather abruptly during the 1984 Summer Olympics in Los Angeles, when the nightly TV broadcast ended with a message from a local station: "Environmentalists killing sheep on Santa Cruz. Story at 11."

"Then it started," said Murray.

## Pig Fight

It was the beginning of a pattern, of scathing letters and damning editorials and accusations that would erupt here and there and wherever the bullets and poisons started flying in the name of conservation. In the mid-1990s the Nature Conservancy found itself again the subject of scorn, this time for killing pigs in defense of Hawaiian rain forests. The pigs—feral, bristly hulks descended from barnyard stock—had taken to rototilling great swaths of the fragile Hawaiian flora, and killing the rarest of birds in innocently diabolical ways. The pits left by the pigs' rootings and wallowings filled with rain, which made ideal breeding grounds for swarms of foreign mosquitoes. And soon

Hawaii's native songbirds—some of them down to handfuls of survivors—began dying slow deaths of avian malaria. Neither fences nor the nascent science of feral pig sterilization offered any practical answer to an enemy so firmly entrenched in such impenetrable mountain wildernesses of razor-spined, rain-drenched jungle. So with more amiable options precluded, the conservancy started shooting and snaring Hawaii's feral pigs in its preserves.

Snared pigs did not often die nicely, and People for the Ethical Treatment of Animals, a national animal rights organization, fashioned a national publicity campaign around that fact, posting graphic advertisements of strangling pigs suffering festering gashes. Failing traction in Hawaii, PETA went for maximum publicity at the conservancy's high-rise headquarters in Arlington, Virginia, across the Potomac from the nation's capital. On the first of a two-day offensive, PETA demonstrators—one of them dressed as a pig—picketed outside the conservancy's offices. The Washington press all but ignored them. Conservancy employees spent their lunch hours enjoying the show from the skywalk. On the second day PETA got impatient, and things got physical. A bucket of blood-red paint splashed upon the doorstep. Protesters stormed the building. Ray Culter, the conservancy's director of administration, found himself wrestling with a big, pink, bipedal pig. Culter sent the pig fleeing, minus his head. PETA's remaining troops mounted a last-ditch offensive, lying in a chain across four lanes of city traffic, which brought the handcuffs and paddywagon and an end to the fracas. The conservancy went on killing pigs in defense of its Hawaiian sanctuaries, while back at headquarters, said Culter afterward, "we kept the pig head."

Years later the Nature Conservancy found itself facing yet another pig fight, this one back on the familiar contentious territory of Santa Cruz Island, but this one adding a few new twists, featuring a bizarre little fox and an unlikely new adversary.

The fox in question was a diminutive island specialty the size of a large kitten with a trusting temperament to match. Numbers of the Santa Cruz island fox in the 1990s had free-fallen from two thousand to less than one hundred, its rescue posing a Solomon-esque dilemma. The fox's demise followed a roundabout route back to the feral pigs that had assumed command of Santa Cruz—pigs that had come to endanger the fox by way of a more prob-lematic accomplice, the golden eagle. Immigrant eagles, enticed from the mainland by the allure of squealing piglets, had taken to snatching the clueless little foxes like candy.

The Conservancy and the Park Service thus found them-selves at odds with two invaders, one of them a wilderness icon. And as long as eagles remained, the fox would be in trouble. (Eagle trappers had discovered the remains of thirteen foxes in one nest.) It was entirely possible, and even frightfully likely according to one scientific model, that to remove the pigs with-out first removing the eagles would have the raptors descending with undivided attentions upon the little foxes, to their cata-strophic end.

For the sake of the Santa Cruz island fox, both eagle and pig had to go. In 1999 local raptor specialists began trapping golden eagles and hauling them back to the mainland, to some minor grousing from the critics. In 2005 Prohunt, a squad of profes-sional eradicators from New Zealand, with guns, traps, hunting dogs, helicopter sharpshooters, and hormonally juiced Judas pigs, began systematically routing and gunning every last pig off the island. And the crowds, as they say, went wild. KILLING SPREE OFF OUR COAST, blared a headline. ISLAND PIG ERADICATION SPURS WILD CONTROVERSY, blared another. Letters to the editor dripped with venom; old arguments and familiar opponents resurfaced on Internet posts: "The pigs have been demonized and accused of imaginary crimes," opined Rob Puddicombe, the onetime accused saboteur of the Anacapa rat poisoning. "The same arrogant assumption of superior mentality that brought us Three-mile Island, Vietnam, Wounded Knee and Waco is alive and well at Channel Islands National Park."

Fifteen months later, under budget and more than a year ahead of schedule, Prohunt dispatched the last of 5,036 pigs with a high-powered rifle. The foxes had survived the pigs, the eagles, and the cross fire. The sniping from the mainland quieted. Within three years, Santa Cruz was bounding with more than seven hundred foxes, reproducing more like rabbits.

Saving the Santa Cruz Island fox by lethal means, for all its eventual vindications, came with a caution for the entire budding profession of conservation eradicators. This business of rearranging ecosystems was riddled with hidden strings and trip wires, and saddled with that eternally haunting price for mistakes and miscalculations, a price often measured in lives.

## Expect the Unexpected

Thus toughened to the task, the Nature Conservancy signed on as the third axis in the alliance to take the rat out of Rat Island. To lead the operation, Gregg Howald chose his Island Conservation teammate Stacey Buckelew, a student of seabirds from the Antarctic to the Aleutians and a recent veteran of the Anacapa campaign. Buckelew in turn was surrounded with a team of advisers: In addition to Howald she was matched with Steve Ebbert, supervisor of the refuge's fox-eradication program, Steve MacLean, director of the Nature Conservancy's Bering Sea program, and New Zealand's Pete McClelland, reigning world leader of high-latitude eradications.

The protocol by now had become fairly well established, and impressively so of late. McClelland's clearing of Campbell Island topped a growing list of some three hundred island rodent eradications on the books, from the tropics to the high latitudes of both hemispheres. The proven approach for wiping rats from wilderness islands had settled upon a few basic principles: Deliver a lethal dose of poison bait, preferably the anticoagulant brodifacoum, to the nose of every rat on the island in their hungriest of times—or at close as one could safely get. In Rat Island's case, that time would be October, when the brief flush of summer greenery and nesting birdlife gave way to the long, raging siege of the Aleutian winter.

After four years of planning, permit seeking, meetings, and conversations across the continents, the Rat Island operation was cleared to commence. On September 17, 2008, the merchant vessel *Reliance*, a 160-foot converted crabbing boat out of Seattle, loaded with fifty tons of rat bait, five thousand gallons of jet fuel, and another six tons of camp shelters, food, and equipment, shipped out of Homer, Alaska, heading west on a weeklong, thirteen-hundred-mile journey across the Aleutian archipelago to the staging harbor of Adak's Sweeper Cove.

Two days later, two Bell Long Ranger helicopters followed, each carrying two of the best pilots in their respective hemispheres. The Bering Sea, true to her nature, soon met them with a wall of fog and winds blowing sixty knots. Alaskan pilots Mike Fell and Merlin Handley dropped down, looking for a window of visibility. They found themselves a hundred feet above the water, pinched by fog above, high seas below, buffeted in a tunnel of turbulence. The storm fought them across the sea. They island-hopped from one fuel cache to the next, to the villages of King Salmon, Port Moller, Cold Bay, and Dutch Harbor, where they rested for the night. The next day, more of the same, more winds and rain and fog and white knuckles on the flight stick, to the Islands of the Four Mountains, to Atka, and finally putting down with an exhalation, at Adak. "It was not a trip for the faint of heart," said the Aleutian veteran Fell. "I would not have wanted to be a rookie out there."

On the third morning Adak treated them to more of the same, Fell anchoring his million-dollar machines to keep them from blowing into the Bering Sea. Into relenting winds, the helicopters finally lifted off, heading another two hundred miles west, where the *Reliance* was now maneuvering into position. One more stop, for a drum of fuel cached on a beach in the Delarofs, and sixty miles later the pilots were at last hovering down over a patch of tundra on Rat Island.

The crews thus assembled on September 26, as bees to a hive. The helicopters went into service ferrying cargo from boat to shore, sling-loading the ninety-one drums of jet fuel, the 220 gallons of gas for generators, the 330 gallons of kerosene for heaters, the weatherport shelters, the generators and kerosene heaters and propane stoves, the inflatable skiffs and outboard motors, the survival suits, the boxes of food, and the boxes of bait and rat-traps. From the empty spaces of Rat Island's lonely tundra sprang a tent city.

The baiting was set to commence on September 28, weather permitting. Every step of the invasion came cushioned with layers of contingency plans, nearly all of it based on the promise of delays. But on the morning of September 28, as if the gods had suddenly tired of a prolonged prank, the foul weather on Rat Island broke. Buckelew and the pilots gathered at dawn, as planned, to talk weather, to consult the forecasts and decide whether to risk the flight. There was little to discuss. The radio and the readouts had nothing but clear sailing to report.

Buckelew briefed the field crews, reviewed assignments, and

checked the radios. The New Zealand aces Graeme Gale and Peter Garden set their rotors spinning and brought their helicopters to hover over five men in hard hats, who began loading fifty-pound bags of brodifacoum-laced pellets of rat chow into the dispensers.

With seven hundred pounds of poisonous payload slung underneath, a hand signal and a radio call sent them off. Garden and Gale flew their lines, eyes darting between instrument panel, bait bucket, compass, map, and ground, flying their perfect paths, meticulously sowing their swaths of poison. The two had been trained to fly straight lines in hurricane crosswinds, all the while aiming precise doses of pellets from a half-ton bucket. They covered the cliffs with side-glancing sprays, sticking pellets to ledges on vertical rock. They painted by number, as it were, at the rate of nearly a ton of bait per hour, as steadily as only a few humans on Earth could.

They divided the island into thirds, moving from one block to the next, each block ostensibly small enough to cover in one day. They spent extra time saturating the beaches—the rats' prime habitat—double-dosing the island perimeter. It was a move calculated to reach the richest concentration of rats, a move that would later bring trouble.

Once the pilots had brushed with their broad strokes, ground crews applied the finishing touches. Around the lakes they followed with buckets of bait, marching in lines thirty feet apart, sowing in unison by the measured handful. All aimed with nothing less than perfection as their goal, with the understanding that one rat left untended could scuttle the entire mission.

Buckelew's team had planned for the Aleutians' typical schizophrenic weather pattern, a Jekyll and Hyde performance lasting six to ten days and generally split between spells of pleasantly light winds and misty clouds and Siberian fronts of relentless rains and hurricane-force gales. Hopes were to hustle and cover the island during any wind-free windows, hunker down during the ensuing blow, then spring forth with the earliest lull. But by some heavenly intervention, the Aleutians' angelic alter ego held sway.

The pilots took full advantage. They finished their first sweep, covering the island without incident. The weather, confounding all expectations, held calm. The radio reports repeated a monotonous forecast of good weather holding. Finally came news of an ominous system heading their way. It never showed up. The crews waited a day to let the bait do its work. And still the fair weather held. Vernon Byrd, the refuge's chief scientist and thirty-seven-year veteran of the Aleutians, had never seen such a spell on the cusp of the Bering Sea's stormy season. "Somebody was looking out for us," he said shortly after. It was a blessing that he would later look back on with second thoughts.

With the window of good weather holding, the chiefs met and decided to take the gift and run, moving quickly into the second spraying. Crews were briefed and mobilized. Again the helicopters lifted off, hefting their buckets of rat chow, again showering the island back and forth, mountain first, doubling up on the coastline, ground crews hand-feeding the bait to the lakesides. Three days later, more than a month short of the six-week siege Buckelew had supplied for, Rat Island was apparently finished.

About that time, a printer on board the *Tiĝlax̂*, now anchored offshore for support, extruded the latest meteorological report. The chart drew immediate attention. The system heading Rat Island's way was depicted in isobars of atmospheric pressure, one stacked upon another like the topographic contours of a cliff. Over that cliff of pressure the winds were hurtling. Even by Aleutian standards, this was going to be a dandy.

Billy Pepper, captain of the *Tiĝlax̂*, radioed the report to shore. The demobilization drill immediately commenced. Weatherports and pup tents were disassembled, garbage was bagged, food boxed, radios packed, fuel drums and bait pods and pallets airlifted to the deck of the *Tiĝlax̂*. After their last delivery the helicopters lifted off once more, with all pilots and Buckelew aboard, and fled east to beat the storm to the mainland. The *Tiĝlax̂*, loading the rest of the crew and stacking the last of the cargo six feet high on decks fore and aft, pulled anchor and sailed for Adak.

Twenty-one hours later Pepper brought the *Tiĝlax̂* to dock in Sweeper Cove, hurriedly unloaded all but the ship's crew, and, with the decks still precariously piled high, headed directly back out for home port in Homer. Soon after he sailed, the sheltered waters of Sweeper Cove were rising on nine-foot swells and one-hundred-mile-per-hour winds. Those booked for the biweekly flight out of Adak would be stranded for two more days, until aircraft could once again land the island. The *Tiĝlax̂* got as far as the Islands of the Four Mountains before the gales caught up. And for the final seven hundred miles, boat and crew tossed through twenty-foot seas, running for cover in the lee of islands, playing cat and mouse with the storm. Seven days after leaving Rat Island, the *Tiĝlax̂* pulled into port in Homer, Pepper and crew chalking up another memorably nasty week on a typically nasty sea. "That was a terrible piece of ocean," said Pepper, "but the crew was good at lashing. We didn't lose a stick of cargo."

## Be Prepared to Be Surprised

With the *Tiĝlax̂*'s safe return, the real work was done. The team could now only watch and wait, for one of two scenarios: In the first, seven years and more than two million dollars worth of planning and proposing—not to mention the chance to save Kiska—had just slid down a rat hole. In the second, which most on the team believed more likely there would soon be nothing but new birds and greener hillsides to report from Rat Island. "Everything went incredibly smoothly during the operation," said McClelland. "I would be amazed and hugely disappointed if the eradication hasn't succeeded."

There was little reason to doubt that by then the rats of Rat Island were already dropping in droves, with any survivors soon to follow. The weather had been as perfect as during any eleven-day stretch in any Aleutian October of collective memory. There was no reason to think that the latest rat eradication had failed.

"Rat Island has put island restoration through invasive-vertebrate control on the global scale," said Howald. "It's basically ripped that world open now. Be prepared to be surprised by what we find on Rat Island."

The following April, six months after the poison had been laid, a crew sailed back to Rat Island for the first checkup since its surgery. The searchers walked the perimeter, looking for signs of rats, surveying for birds, hoping to mark the awakenings of life on the ratless shores of Rat Island. Before long somebody did spot a bird. It was the corpse of a glaucous-winged gull, dead by what means, nobody could say. They recorded the gull and walked on. Soon came another dead gull. And another. Somebody came upon a huge dark carcass with the massive talons and saber beak of a bald eagle. And then another. What might otherwise have been more simply blamed on an act of nature—the punishing Aleutian winters were notorious for weeding the young and the weak—now came with added complications. For poison had now entered the picture.

On the beach, more gulls, more bald eagles. The more people

walked, the more dead birds for the data sheets. There were forty-one eagles, 173 gulls, and counting. Buckelew began sending the birds' livers to a toxicology lab in the National Wildlife Health Center's laboratory in Madison, Wisconsin.

While all waited on the test results, there remained hopes that these kills were somehow within the Aleutians' bell curve of normality. Howald, who knew brodifacoum better than anybody else on the team, resisted the easiest and most damning conclusion: "I don't know that they're nontarget kills yet. We merely noticed a higher number of dead birds on the beaches. We've also been finding pelagic seabirds, where there's zero to no chance of rodenticide in them. I don't know what's going on yet. Certainly the number of eagles presents the biggest concern."

Howald tried to think it through. He recalled his study on the poisoning of Langara, where there had been many eagles before and roughly the same number afterward. There hadn't even been forty-one eagles living on Rat Island. How could so many have wound up dead?

"Certainly birds have to die somewhere," he said. "It's not uncommon to find dead birds. We're putting the island under a microscope. People don't typically do that under normal reconnaissance. There have been incidents of numerous carcasses found in the Aleutians for which we don't know the causes, but they're certainly not poisoning. It does happen that starvation is not uncommon for birds of prey. Basically their energy requirements are so high . . ."

The Rat Island eradicators, at turns proud, at turns defensive, were ultimately a bit confused with their public message. Even as the bodies and suspicions were stacking up, the Nature Conservancy was posting glowing news of the eradication's success, with slide shows on its Web site celebrating a newly hatched oystercatcher in its puff of black down. "Black oyster-catcher nests such as this one discovered during the 2009 summer field season are the first ever recorded on the island," read the caption.

Well, maybe. The fact was, oystercatchers had been known to nest on Rat Island before the rats were removed. The Web site of Island Conservation also shared the good news, albeit with another convenient dash of omission: "No sign of invasive rats was found on the island and several bird species, including Aleutian Cackling Geese, Rock Ptarmigan, Peregrine Falcons, and Black Oystercatchers were found nesting!"

Well, yes, such birds were found nesting, just as they had been found nesting all along on Rat Island, at least since the foxes had been killed off years before. The congratulations came capped with a touching thirty-second video of two downy oystercatchers hatching on the rat-free shores of Rat Island. Through it all, nary a word of dead gulls or eagles.

As every biologist on the project knew, however, it was in fact too early for great expectations in an ecosystem likely to be years on the mend. On June 11 the three partners in the Rat Island eradication released a more forthcoming statement to the press, this time including the discomforting details.

"Biologists," it stated, "have found 157 juvenile and 29 adult glaucous-winged gull carcasses and a total of 41 bald eagle carcasses that appear to have died in recent months . . . While some level of winter die-off of these species is not unusual on islands in the Aleutians, and avian die-offs are not uncommon in Alaska, these numbers are cause for concern and further investigation. The Service is very concerned by these levels of mortality and is doing everything possible to expeditiously determine the cause of death."

All wishful thinking was soon enough dashed when in late June the test results that everybody had feared came back. Every liver sampled—from two bald eagles, two glaucous-winged gulls, one peregrine falcon, and one rock sandpiper—had tested positive for brodifacoum toxicosis. This was no longer a natural weeding from a hard Aleutian winter. This was a rescue that had gone awry.

This was also now a legal issue. Many of the victims, after all, were not just any bird, but the nation's symbol, a protected bird. Searchers were sent back to scour the island for more carcasses. A federal investigator was among them. Howald, Buckelew, and the refuge's eradication specialist Steve Ebbert were questioned about what had happened to those forty-odd eagles on Rat Island and why. The partnership commissioned a panel of three independent reviewers from the Ornithological Council, based in Washington, D.C., to examine the eradicators' protocol and assumptions that might have led to the demise of more than four hundred unintended victims on Rat Island.

Much was already apparent. Assumptions, many of them based on the experience at Campbell, had been wrong. Said Buckelew, "We did two bait applications. Campbell had only one. The risk was higher, but what was the trade-off? What if we fail? The recommendation was, if you have the resources to do two applications, do them. The bait didn't move as quickly as we calibrated. It didn't break down. I don't think we have the same microbes, the same temperature as Campbell."

There was also a question of timing. Why, for example, was nobody around to witness the first signs of trouble? Why, with more than six weeks of supplies, did the entire expedition need to decamp after eleven days? The anomalously splendid weather that had allowed the baiters their rapid deployment and subsequent exit, had perhaps allowed them to miss the first sick gull. Perhaps with more patience, they might have noticed the first gathering of eagles.

The number of eagles that had found their way to Rat Island had surprised everyone. There had only been four or five pairs known to be nesting there. Somehow, after the poisoning, word had gotten out among the Aleutian eagle network that there were dead or dying gulls and rats for the taking on Rat Island. And in they flocked, to their deaths.

Which, for certain spectators, seemed not so bad a deal after all. Alaska was a state where both the gulls and the bald eagles had become populous to the point of nuisance, their numbers inflated by garbage and offal from the Bering Sea's two-billion-dollar-a-year fishing industry. The eagles eventually tallied on Rat Island were forty-three of some twenty-five hundred inhabiting the Aleutians.

"To be honest, they're trash birds," said pilot Fell. "We've been out here a long time, and I can tell you there are more than plenty of eagles in the Aleutians."

"The nontarget issue is frustrating," said Art Sowls. "But if you have cancer, you have to decide if you're going to have chemo. Truth is, if they get all the rats, all the nontargets will be better off in the long run. We knew there was the probability of some loss in the process."

"The losses are short-lived compared to restoring the ecology of the island," said Ed Bailey of the U.S. Fish and Wildlife Service. "So you lose a few eagles along the way. Here in Homer they're causing all kinds of grief. They're killing waterfowl, killing sandhill cranes, people's pets and poultry. Eagles in Homer have become trash birds."

"The only reason some people are concerned is because of the eagles," said the refuge's Jeff Williams. "Is that any different than songbirds getting killed? It's a big giant bird, a national symbol, with more meaning and cachet. But everyone knew that. Most people don't know that bald eagles have a very high mortality rate of their young each year. Whether all of those birds that were poisoned would have survived otherwise, I don't know. Will it make it more difficult for permitting future projects, I don't know. But who's going to be the one to complain? The local inhabitants most intimately linked with wildlife in this region think it's unfortunate but not that big a deal."

"They fluffed some feathers, but the bottom line is, they got the rats," said the *Tiĝlax̂*'s Pepper. "If eagles are there in the year 2020, and seabirds are there in 2020, then the end justified the means. The bottom line was, that eagle was going to die anyway. Just like all of us. Whether it was twenty-two or eighteen grams of poison per acre, forty-three eagles, however many gulls, they're all just numbers. I'm a layman. To me, the bottom line is, are there any more rats? If the answer is no, they'll all be back there—all the eagles, gulls, seabirds."

"We're moving forward," said Howald. "Everybody acknowledges this was an unfortunate incident. What can we learn from it? How can we maintain the ability to do rat eradications in the Aleutians and further minimize the risk for nontargets? From my perspective, I'm surprised by the number of birds, but not necessarily the risk for individual species identified."

## Fallen Byrd

Amid the explanations and excuses, shrugs and regrets, one person among the many responsible all but demanded the blame. "You've found the guy whose fault it is," said Vernon Byrd.

Nobody understood the stakes of Rat Island's eradication better than Byrd, and nobody seemed to take the losses harder than he. Byrd had committed his life to protecting the Aleutians. He'd first come to the islands some forty years before, as a navy junior officer stationed at Adak. Adak—Aleut for "birthplace of the winds"—was a place from which those of unraveling minds were regularly shipped stateside for evaluation. At the end of his tour, when most of his comrades were clawing to escape their sentence in this frozen hell, Byrd hurriedly signed on for another go—earning himself a visit from the navy psychiatrist. "That was very unusual for a single person to offer to extend his tour at Adak," said Byrd.

When Byrd finally quit the navy in 1971, he took up the next day as biologist for the U.S. Fish and Wildlife Service in Alaska. He followed in the fashion of his dory-sailing mentor Sea Otter Jones, gravitating to the majestic harshness of the Aleutian wilderness, banding eagles, counting waterfowl and seabirds, and coming to recognize the consequences of the Aleutians' invasion. Byrd could judge from the look and smell of an island whether foxes and rats held sway there. He would see the foxes disappear down the line under the determined chases of Jones and Ed Bailey, and witness the flocks coming back. And when the early rumors started to swirl of something wrong on Kiska, he himself went and witnessed one of the first caches of auklets on the rocks at Sirius Point. He had been as eager as anyone to see Rat Island free of rats and full again of seabirds, and as eager for its success to begin paving the way toward Kiska. And he, along with everyone else who had signed on to take Rat Island back, had badly miscalculated the ultimate cost. But it was Byrd who took his own sword to chest.

"I'm the guy who's supposed to know the most about birds in the Aleutians, and I clearly thought the eagles were not going to be a problem, and I was dead wrong," he said. "You don't have to look any further. I know who should have known best about that."

In June 2010 the crews returned to Rat Island to set up their summer camp, to check again with their hopes of rats missing and birds returning. The carcasses had stopped coming, the poison long gone, all bodies by now dust and feather. The review panel was still reviewing. For now, it was a time of watching and waiting on Rat Island, for the tiniest hint of what was sure to be a long recovery. After a week on the island, however, project leader Buckelew returned to the *Tiĝlax̂* all but bubbling. "The first beach I walked on, the first bird I saw was a song sparrow. I've been on the island four seasons and never seen a song sparrow."

Even Byrd allowed himself a proud moment. "It's looking likely all the rats were eradicated," he said. "Ultimately Rat Island is going to be great. The eagles will come back. The gulls haven't even been affected at the population level. In fact there's going to be a lot more gulls in the long term after the seabirds build back up and gulls have more prey. The ecosystem will be restored. The long term is very positive."

## Chapter 12

## WHITHER KISKA

Amid all the cautious celebrations of Rat Island's costly eradication, there remained the question of Kiska. Once a beacon of distress looming large on the horizon, Kiska suddenly seemed fainter and farther adrift. The urgency in those giant piles of tiny dead auklets that had triggered the Aleutian rat campaign had given way to an Aleutian case of cold feet.

"We continue our commitment to protecting important seabird habitat in the Aleutians," said the Nature Conservancy's Steve MacLean, "but extraordinarily high costs and problematic fundraising have precluded another eradication on a larger island. We are hopeful that we will be able to continue this program, but are concentrating now on learning as much as we can from the Rat Island project."

"Kiska is going to force a fundamental shift in the way we think of doing these types of projects," said Gregg Howald. "Is there stomach for the nontarget loss on the scale of Rat Island? I don't know."

"At this point I don't think anybody would want to take Kiska on," said Vernon Byrd. "The size itself is enough to be completely daunting. The cost will be astronomical. There are still caves on Kiska dug by the Japanese in World War II—scary places, some of them booby-trapped. There may be rats living down there that never see the light of day. I don't want to say Kiska is not ever doable. Ultimately I would love to see Kiska rat free. But I would want at least one more island under our belt before attempting it."

There was every good practical reason for backpedaling on Kiska, not to mention the psychological weight of Rat Island's forty-three dead eagles still heavy on the conscience. By the crudest of calculations, Kiska would require ten times the effort of Rat Island—twenty helicopters, five hundred tons of rat bait, ten ships, and so forth—in an economic climate whose bubble had lately burst. Kiska had streams running with salmon, with scores of bald eagles regularly converging on them, and the potential for collateral casualties to easily eclipse the body count on Rat Island. And how, after all, was one to deal with Sirius Point, to deliver a lethal package to every rat's address in the unreachable bowels of their underground Gotham City?

There remained yet a more baffling reason that the burning concerns for Kiska had cooled. In the years following the least auklets' electrifying collapse of 2001 and 2002, Ian Jones and his students had come back from their summers at Sirius Point with their surveys suggesting something oddly leaning toward . . . normalcy. Young auklets were fledging, in some years, at healthy rates. The rats' grisly caches had grown sporadic. It seemed that the worst of the siege had abated, that the rats had somehow been knocked back.

"When I first went to Kiska there was rat shit everywhere," said Jones after coming off the island in 2010. "Now it's like they're not even there. We would find these carcasses of auklets with holes in their head where the brains were cleaned out. We'd find rats in their burrows with *their* brains eaten out. They were eating each other. Now, *poof!* We don't know what's happened with the rats on Kiska."

Not that the auklets had quite escaped the crosshairs. Those harbored at Kiska accounted for one of only nine tenuous colonies in the Aleutians (a number that had recently been reduced by one when, on August 7, 2008, a surprise volcanic eruption on the island of Kasatochi had entombed some forty thousand chicks under a hundred feet of ash and boulder, leaving their parents and another quarter million breeding auklets looking for new homes). The colony at Kiska was the biggest of an exclusive few, and that standing carried special degrees of promise and peril.

One hazard came self-inflicted. A bird of habit, the least auklet faithfully staked out its breeding lots on slopes of bare boulders, seeking clear views of airborne predators and proper stages for its courting dances. It shunned encroaching plants and obscuring greenery, ironically of the very sort that its own guano tended to fertilize. There would naturally come a time when, with boulder fields fading beneath the foliage, the birds would abandon and the colony fall silent.

Kiska was different. Kiska's working volcano was in the habit of occasionally throwing up new habitat, the signature boulder field at Sirius Point most recently enlarged by an eruption in the 1960s. While other colonies withered with age, Kiska's periodically freshened itself with volcanic face-lifts. Sirius Point offered the auklet's most enduringly alluring real estate in the Aleutians.

Yet even that distinction brought mixed blessings. Jones suspected that auklets and other Aleutian seabirds out prospecting for new homes were not uncommonly enticed by the avian commotion at Sirius Point to stop by and inquire about vacancies. But to what end, with rats still at large, Jones could only darkly guess. "Maybe it's like a hotel where murderers are killing all the people checking in."

Even for Kiska's auklets in residence, the latest stay of execution was by definition only temporary. "It's certainly possible that rats will boom again. We could have a run of go-go years for rats that could kill the auklets," said Jones. "It's a war of attrition. The end point is unknown. But as long as there are rats on this island, that colony is in danger."

## THE ENDLESS FLOCK

All that anybody could say for sure, as of June 2010, when the *Tiĝlax̂* made one of its seasonal sweeps past Sirius Point, was that the auklets of Kiska were still performing a show for the ages. The *Tiĝlax̂* had arrived at the dusky hour of half past ten, and as was customary whenever the vessel happened upon this special place and time in the world, the engines had come to an idle and the boat to a slow drift. The birds were returning from the sea, skeins of auklets skimming over the water, clouds of auklets billowing over the far horizons. And on they came with an ever-frenzying pace and the musical roar of their multitudes, ascending the snowy heights of the Kiska volcano.

It was impossible to say whether the torrents of life raining upon the headlands of Sirius Point represented more the indomitable force or the fading remnant of a far richer storm of birds. Nor could anyone say what the rats' next move would be. Maybe they had met their match in the culling winters of Kiska. Or perhaps this was the season that would find them again tearing through the auklets. Given its global record of conquest, there was no betting against the rat in the long run.

But for the moment, one could only stand dumbstruck before the mind-bending enormity of the auklets' masses, as one pondering the brink of the Grand Canyon. The deeper the gaze, the dizzier the reckoning of scale. For every flock of birds there was another behind it, and another behind that, repeating to the end of sight. Witnesses had sometimes compared the phenomenon of Sirius Point to the northern forests' legendary flocks of passenger pigeons, obscuring skies for hours in passing. It was perhaps of no trivial portent that the passenger pigeon—slaughtered en masse for fertilizer and hog feed—fell from its untouchable flocks of billions in the mid-1800s to exactly zero in 1914.

An hour into the show, Kiska's auklets were still sweeping endlessly from the sea, swarming and swirling over the point. The diesels of the *Tiĝlax̂* rumbled to life, the ship moved on, and Sirius Point melted into the horizon, beneath glowing heavens still streaked with fleeting wisps of living smoke.

*Epilogue*

# ISLAND EARTH

L ATE IN AUGUST of 2010, nearly two years after the poison had been laid, the last crew of the season came off Rat Island after the final and deciding survey, with a glorious lot of nothing to report. No baits chewed, no snap traps sprung, no rat prints or scats in the sand. "Rat Island is officially declared rat-free!" exclaimed a press release from Island Conservation.

That summer oystercatchers had raised chicks on the rocky shores of their ratless new island, song sparrows had sung from the same spaces found empty not long ago. The gulls had gathered by the score about the wrack lines, and eagles had returned to nest as if nothing had happened.

There was no word in the announcement about dead birds, but much about living ones, about the sparrows and oystercatchers and gulls, the pigeon guillemots, rock sandpipers, common eiders, red-faced cormorants, and gray-crowned rosy finches, all confirmed to be nesting on the island. "Restoring habitat on Rat Island for native seabirds is the most ambitious island habitat restoration project ever undertaken in the Northern Hemisphere and the first in Alaska," continued the release. "Thanks to everyone who supported this incredible conservation achievement."

The apparent stirrings of resurrection—still too early for science to confirm, though not beyond the devoted to proclaim—suggested images of a Rat Island already on the mend. Hype and hopes notwithstanding, chances were good now, and certainly better than they had been in two hundred years, that there would soon be a new commotion of life to Rat Island, like the sounds of a city street awakening with the first glow of dawn.

There will likely come a day, perhaps in a few years yet (perhaps it has already happened), when the odd prospecting puffin or storm petrel will take a chance and land on the suspiciously uncrowded shores of Rat Island, to find its fears unfounded. And in time the pioneer's boldness will embolden others, and the clan will grow ever bigger and noisier and more irresistible to passersby. The flocks will multiply, and the headlands will again be busy avenues of birds commuting from the sea. The guano will rain, the hillsides will bloom, the dead will be forgotten.

## WORLD WAR

Even as the crews were coming off Rat Island, others were already heading out on new record-breaking missions. Peter Garden, ace eradication pilot of Rat and Campbell islands, was on his way to the South Atlantic, to destroy the infamous albatross-eating mice of Gough Island. Garden was also slated for duty in the subantarctic seas for what promises to be the next granddaddy of all rat eradications. South Georgia Island is seventy-five miles long, half covered in snow and glacier, and spectacularly brimming with penguins and albatross and a host of other seabirds being catastrophically trimmed by millions of rats—every last one of which Garden and an army of colleagues have designs on killing.

South Georgia's anticipated restoration would add to a world tally of island eradications that has already topped eight hundred. With the many victories, however, have also come a few black eyes. In the fall of 2010, even as the engineers of the Rat Island overkill were anxiously awaiting their review, more bad news was coming from another world-class bastion of wildlife, from the South Sea island of Macquarie. Fifteen years before, Tasmanian wildlife officials had begun systematically killing cats on Macquarie, bagging the last one in the year 2000. Seabirds began recovering, only to suffer from ecological whiplash. The missing cats were replaced by a plague of some 130,000 rabbits and a corresponding flush of rats and mice, eating the island tussocks to nubs, invading the bird colonies, inviting new disaster. In 2006, the rabbit-scoured hillsides gave way under heavy spring rains, emtombing untold numbers of nesting penguins below and triggering an international landslide of finger-pointing press, the bulk of it deriding the debacle as an example of man's ham-handed tampering with nature.

In 2010 Macquarie's managers went back to finish the job, bombing the rabbits and rodents with brodifacoum. Horrible weather intervened, the eradication failed, and more than four hundred birds were found inadvertently poisoned—the latter embarrassment again making the headlines.

Undeterred, the eradicators march on. Island Conservation's Karl Campbell—of Galápagos goat-killing fame—and his partner-in-mischief Josh Donlan have been flirting with heretofore unthinkable leaps to those islands once called continents. Campbell and Donlan—who four years ago directed a battle discharging half a million rounds of ammo and killing 160,000 goats on a Galápagos island the size of San Francisco—have recently been visiting with the governments of Chile and Argentina, with talk of saving fifty-four thousand square miles of forest in Tierra del Fuego, now flooded and endangered by dam-building beavers from North America. "You can see their lakes from an airliner," said Donlan. "It wouldn't be cheap, but the alternative is beavers making their way to Patagonia."

Campbell and Donlan's exploratory forays, onto an archipelago better approximating the mainland, give hint of the island campaign's inevitable destination. It should go without saying that the plague of biological invasions did not begin or end on the shores of oceanic islands. "We must make no mistake; we are seeing one of the great historical convulsions in the world's fauna and flora," wrote Charles Elton in his 1957 book, *The Ecology of Invasions by Animals and Plants*. Elton was a founding and visionary ecologist from England who drew the first and most lasting portrait of a world eating itself alive. "We are living in a period of the world's history when the mingling of thousands of kinds of organisms from different parts of the world is setting up terrific dislocations in nature. We are seeing huge changes in the natural population balance of the world."

Elton wrote of African mosquitoes inadvertently shipped to Brazil, igniting "one of the worst epidemics Brazil has ever known." During the disaster, also, "hundreds of thousands of people were ill, some twenty thousands are believed to have died, and the life of the countryside was partially paralysed." Elton charted the wreckage of the chestnut blight, a fungus from Asia that spread like fire to the ecological ruin of the dominant tree of the eastern U.S. forest. He foresaw the dangers of the sea lamprey sneaking through the gates of the Erie Canal, on its way to extinguishing three native species of Great Lakes fish; he presaged the ecological catastrophe of the European zebra mussel, ferried to U.S. waters in the ballast tanks of transatlantic cargo ships and now clogging pipes and smothering life, threatening dozens of native mussels, and costing billions of dollars to fight it.

The world has not gotten any safer since Elton first sounded the alarm. The new global village trades not only in electronics and soybeans but also in weeds, disease, insect pests, and more of the familiar cast of misplaced mammals—an epidemic of epidemics. In the United States alone, some fifty thousand alien invaders have been helped ashore and across the borders. The invaders include more than a billion rats and one hundred million house cats, the latter of which have been implicated in the demise of perhaps a billion small mammals and lizards and birds every year. U.S. forests are serving as barnyards and feedlots for four million feral pigs. One economic accounting of the invaders' damages comes to $120 billion per year, with a discomforting caveat: "If we had been able to assess monetary values to species extinctions and losses in biodiversity, ecosystem services, and aesthetics," wrote David Pimentel and his research colleagues, "the costs of destructive alien invasive species would undoubtedly be several times higher."

Americans have been slow to concern themselves with these figures, certainly if stray cats are any measure. In 2008, in Galveston, Texas, a man who shot one feral cat as it was chasing a rare shorebird was arrested and faced up to two years in jail and a sixty-thousand-dollar fine before the charges were overturned. There are now organizations throughout the country actually promoting and feeding feral cat colonies. They are operating very successfully in Hawaii, the U.S. capital of extinction and endangerment, where some of the rarest birds on the planet are still being taken by the subsidized cats.

New Zealanders, on the other hand, have taken the practice of killing for conservation to the level of civic duty. In the same way that a coterie of Audubon Society members might gather with their binoculars for a Saturday-morning bird walk through Central Park, citizen groups in New Zealand are now marching into their local woods armed with snap traps. Such is the state of conservation in an island nation whose vanishing native fauna has already been picked half clean.

"Places when I was a kid wandering around the bush, I used to see and hear kokako," said Bruce Thomas, the former battler for Breaksea. "I used to sit out on the ridges and hear kiwi. They don't exist anymore. They're all gone. One day you wake up and say, 'Oh, I haven't heard a gray warbler for a long time.' The bush is going silent."

Thomas is now a freelance conservationist, bent on supplying New Zealand's rat-killing community with the proper tool for the job. In his toolshed he has developed a rattrap that is lighter (a trapper can carry 150 in a backpack), more foolproof (featuring a trigger that prevents even the most thumb-struck handyperson from smashing his own fingers), and ultimately more deadly. "I'm not out to make my first million dollars," said Thomas. "I'm out to take my first million rats."

Thomas's trap kills quickly. Ninety-nine of every one hundred rats tripping one die of a crushed head or broken neck. "They're all head shots," said Thomas. "Death should be instant."

More are agreeing on that. The New Zealand Department of Conservation has over the past decade been replacing its venerable snap traps with new precision designs that kill far more quickly. And three young entrepreneurs from Wellington, calling their little company Goodnature, have built what they believe is an even better rattrap, featuring a plastic tube and a $CO_2$ cartridge that propels a plastic plunger with head-pulverizing force. The device—named the Henry—not only renders rat or stoat or possum instantly dead, but also automatically resets itself to await the next victim. "Killing is part of our culture," said Stu Barr, cofounder of Goodnature. "But it's not the animal's fault they're here. It's our fault really. We've got to treat them humanely, until we kill them."

In the Kiwis' invaded kingdom, the kill trap has become an icon of conservation; the fence is another. One of particularly epic proportions has recently gone up in the center of the country's North Island. The fence—the last link of it erected in 2006—stands eight feet high and thirty miles around, enclosing in steel mesh a twelve-square-mile mountain island of forest, what its builders hope will someday become a sanctuary of primeval New Zealand. The Maungatautari Ecological Island is financed and run by a citizens' trust, whose written aim is no less than "to restore the dawn chorus, to fill the forest to capacity with native birds, insects, reptiles, frogs and other wildlife, and to share it with all New Zealanders."

As prerequisite, managers of Maungatautari have emptied their fenced forest of all mammalian invaders. Their next step is to reassemble the forest's conspicuously missing pieces, its kiwi and kokako, its giant weta and tuatara. And the ultimate prize now being quietly considered for transfer to the confines of Maungatautari is the reigning icon of New Zealand's fight for life, a bird once believed to be extinct.

## Long Walk Home

At the turn of the second millennium, in the years following the kakapo's terminal diagnosis and subsequent admission to intensive care, the patient miraculously rallied. Following that bumper crop of rimu nuts on Codfish Island in the summer of 2001, the kakapo responded with a crescendo of booming and mating, and the next spring with a bumper crop of chicks. Twenty out of twenty-one female kakapos on Codfish mated. And each, of course, had a midwife at her side. As the nesting season hit high gear, Don Merton and crew matched pace through the nights, candling eggs, nursing chicks, removing infertile eggs with hopes of new layings. By the time the rush of 2002 was through, an army of exhausted nest-minders had welcomed another twenty-four kakapos into the fold, jumping the world population to the dizzying sum of eighty-six birds.

The years following were spent waiting for another like 2002. So much depended on the fickle fruiting schedule of the rimu trees, and all was precariously limited to the one island of Codfish—the only place on Earth where the kakapo still boomed and bred. The rimus' next substantial fruiting came in 2008; the kakapos responded with six chicks. More rimus fruited the following year, and this time the roof blew off. The year 2009 went down as the greatest in kakapo conservation history. Birds mated multiple times; some laid a second clutch. Those needing assistance were artificially inseminated. By the time the books were closed on the kakapo's record-breaking breeding season, the population had vaulted to 124 birds.

But there had developed a dark side to the kakapo's booming fortunes. One of the diseases most feared for the little cooped-up population on Codfish had already begun to show symptoms. Sperm of male kakapos were sprouting extra heads and tails. Eggs were lying infertile. The little band of inbred kakapos was coming apart at the genes.

To no great surprise. All but one kakapo had originated from a tiny remnant of survivors on Stewart Island. The one exception was the lovable old bird from Fiordland named Richard Henry. The kakapo that Merton had pulled from the brink in the Esperance Valley in 1975, the sole carrier of Fiordland blood, had in his thirty years of island life sired all of three offspring, none of which had yet entered the mating game. In the intervening years Richard Henry had shown only sporadic interest in competing with younger males. He was blind in one eye, likely an old battle wound. It was quite possible that the kakapo's knight in moss green armor was more than a century old. Seasons would go by when Richard Henry's hormones no longer responded to the ripening of the rimu. Merton would coddle and coax his beloved kakapo, following him around with treats of apple, enticing him at every turn with offerings of more food, hoping to fatten the grand old parrot for one more go on the courting grounds.

The prospect of rescuing the kakapo on the ebbing virility of Richard Henry Kakapo was but half the worry. Beyond the genetics lay another bottleneck, of space. In running up their numbers, the Codfish kakapos had run themselves out of room. (A single kakapo might command 150 acres of territory, wandering twenty miles in a night.) The island-bound birds had begun to exhibit anxieties. Young males had started breaking into nests and attacking chicks. Codfish had gotten dangerously crowded.

Merton, the man who had once lobbied so relentlessly for taking the kakapos into hand, now found himself campaigning to turn them loose. The time had come for someplace big enough and safe enough to let the nursemaids stand back and let the kakapos prosper on their own.

Such places made for a short list. There was Campbell Island, forty-four square miles of predator-free wilderness far adrift in the Southern Ocean—perhaps too far adrift. There was no telling how the immigrant kakapo would fare in this foreign land, or how the Campbell ecosystem might in turn fare with this bizarre new bird in the mix.

There had also been talk of clearing the predators from Stewart Island, third largest island of New Zealand and a former home of the kakapo. Stewart, though, came with people, some of whom held certain opinions about government helicopters dropping poison from the sky, some of whom believed that their pet cats came before kakapos.

Lately those now in charge of the kakapo's survival have their eyes focused on Fiordland's Resolution Island, where more than a century ago the man Richard Henry attempted the first kakapo rescue. In July 2008 eradication crews returned to vanquish Henry's demon, trapping what they believed to be all 258 stoats over all thirty-one square miles of Resolution. However romantic the notion of the kakapo's returning to Resolution Island, the reality comes with a cold splash. For there remain mice on Resolution. With the mice there remains a constant lure for mainland stoats contemplating the crossing, and with the stoats a commitment of eternal vigilance in guarding the island's shores.

Those now in charge of the kakapo's survival, it should be noted, no longer include Don Merton. After thirty years at the forefront of kakapo research and rescue, Merton retired in 2005. He has since volunteered his services to the kakapo team; the team has not since taken up his offer. For Merton the physical separation from his birds—Richard Henry above all—has weighed heavily. "It is a huge psychological wrench to no longer be intimately involved, after half a lifetime of intense association with this remarkable creature," said Merton. "I regard him almost as one of the family."

Merton would badly love to see Richard Henry back home, if such a place still exists. However much Resolution may now stand as the kakapo's most immediate hope for sanctuary, it ultimately renders the bird an orphaned refugee whose mother country died years ago.

Or maybe not. There remains at least one place that might serve for a homecoming. It survives as a spectacularly towering amphitheater of sheer rock walls high in the wildest reaches of

Fiordland. Sinbad Gully was one of the last two valleys of Fiordland found to harbor kakapo, the place where Merton as a fledgling wildlife officer fifty years earlier began his lifelong search. There is good reason that the Sinbad kakapos held out so long, even as their last bastions fell. Those skyscraping slopes spanned a diversity of life zones that harbored a cornucopia of kakapo food. Those forbidding walls were the last fortress of Fiordland to be stormed by stoats.

Merton believes that Sinbad could be that fortress again. "It's a natural mainland island," he said. Sinbad is surrounded on three sides by sheer cliffs and on the fourth by the sea. The gully is a geological stoat fence. "Stoats are capable of coming down out of the top four to five thousand feet," said Merton, "but it wouldn't happen very often." His idea—and he is not alone in the suggestion—is to clear the valley of predators, bring the kakapo home, and guard the few entrances against reinvasion. Or maybe not.

"It is not high on the current list," answered Mick Clout, chair of the kakapo advisory council, "but it remains an option."

Which is where the fence of Maungatautari comes in. Maungatautari, lying inland, may lack the majestic aura and romantic resonance of Fiordland's Sinbad Gully. But it is nevertheless the mainland, it is native kakapo range, and it is all but waiting. Tentative plans are to bring a few male kakapos to Maungatautari as early as 2011, for a trial visit. Everyone will be watching, to see if the fence holds them, to see if the forest suits them, to listen toward the hills for the beating of a heart.

## One Last Song

With the first of September 2010, on the cusp of another breeding season on Codfish Island, came a report from the kakapo recovery team's lead scientist, Ron Moorhouse. The rimus had fruited well, said Moorhouse. It was shaping up to be another good year for kakapos.

But there was bigger news concerning the castaways of Codfish. Richard Henry Kakapo had been found to be sick. He'd been infected by a protozoan parasite, for how long nobody knew. But it was likely his caretakers had at last discovered the reason he'd been struggling to put on weight, the reason why the kakapo species' most desperately needed father figure had never boomed on Codfish.

The disease, it turned out, was treatable. Kakapo conservationists poisoned the parasites with antibiotics, eradicating every last one of the little invaders from Richard Henry's beleaguered body.

Richard Henry Kakapo had since grown the fattest he'd been in years. At last check, reported Moorhouse, his breast and belly had taken on a spongy feel, the telltale sign of a male kakapo getting ready to boom.

# ACKNOWLEDGMENTS

I owe the idea for this book to Will Murray, longtime sounding board and conservation sage, who alerted me to a radical new wildlife rescue that could "manufacture millions more birds with a swipe." If in my two hundred pages I have conveyed the story half as well as Murray did in his opening five-minute pitch, then I have exceeded my goal.

The financial task of gathering this story was greatly eased by a journalism fellowship from the Alicia Patterson Foundation. Without their help, I might never have witnessed the epic display of auklets at Sirius Point, or peeked into the fantastic hobbit forests of Richard Henry's homeland. To Peggy Engel and the Patterson folks, I remain forever grateful—as I do to my friends and colleagues Josh Donlan, Will Murray, Bill Ripple, and Angie Sosdian, who recommended they choose me.

To the more than one hundred scientists, conservation professionals, trappers, shooters, pilots, and animal rights advocates who shared with me their stories, I am indebted. Among them, I owe special thanks to Stacey Buckelew, Vernon Byrd, Gregg Howald, Helen James, Ian Jones, Lisa Matisoo-Smith, Don Merton, Rowley Taylor, Bernie Tershy, and Bruce Thomas, who each reviewed one or more chapters for their accuracy.

My ventures to certain otherwordly places were made even more memorable by certain special hosts. Captain Billy Pepper and able crew of the M/V *Tiĝlax̂* navigated a mindbending tour of the Aleutians with great skill and matching humor. Jeff Williams and Poppy Benson of the U.S. Fish and Wildlife Service originally helped secure my space on the *Tiĝlax̂*—what has to be one of the most coveted berths on any seagoing vessel. And Lisa Spitler helped make Adak, "Birthplace of the Winds," a most hospitable visit.

In New Zealand, I could not have imagined more generous hosts than Alan and Diane Hay in Auckland, Bruce and Pam Thomas in Nelson, the McClelland family in Invercargill, and the Cumbo clan in Dunedin. And if not for the last-minute heroics and monster truck of Vic and Chandra Vickers during the Blizzard of 2010, I would never have even made my flight out of D.C.

Thanks again to my agent Russ Galen and editor Kathy Belden, who make it seem easier than it ought to be.

Thanks also to Beth and Lori for the writers' retreat in wolf country, and to Dan and Jo of the Hotel Killigrew. To George and Sandy, Bob and Rachael, and Uncle Jon, title consultants and coconspirators of the Proudcastle Breakfast Club. To Thelma, Herm, Dee, and Pam, who found homes for the babies. To Jean, the rat whisperer. And of course, to Kathy, whose selfless love, patience, and prodding lured the resident scribe from his lonely cave to occasional glimpses of sunlight and sanity.

Finally, there remains the acknowledgment I was wishing never to write. Several days after signing off on the epilogue of this story, with the glorious news of Richard Henry Kakapo's miraculous revival, I received a note and photo from Don Merton, who had just returned from visiting the grand old bird at his foster home on Codfish Island. Merton wrote that he had gone to see his beloved friend, after years of separation, to say his farewells. The photo featured a beaming Merton cradling the big green kakapo in his arms. Merton somewhat casually added that both he and Richard Henry were in failing health, and their time would not be long.

The news hit with a hammer's blow. The last I'd heard, Richard Henry had somehow surged, hinting that the most valuable kakapo in the world might be gearing up one last time to step into the ring and vie for the affections of a female. The thought of Richard Henry breeding once again, infusing the little inbred band of survivors with new lifeblood, was enough to make a conservationist's breast swell like a booming kakapo. As for Merton, he'd mentioned earlier in the year that he was still quite fit for kakapo work, should the call ever come. But now here he was, so suddenly saying good-bye.

I wrote Ron Moorhouse, chief scientist of the kakapo recovery program, seeking answers about the mixed messages on Richard Henry's condition. "Don hasn't seen RH in years and RH has aged in that time," answered Moorhouse. "For example he is now completely blind in one eye and moves much more slowly. On the other hand, from my perspective, RH is in much better condition than he was a year ago. Both of us are right, it's just a matter of timescale."

Unfortunately, it was Merton's perspective that proved more prophetic. One month later, I received the news from Moorhouse:

Richard Henry was found dead on 24 December . . . He had recently left his normal home range and we were hopeful he might boom. Looks like the stress of getting into condition for a booming season may have been too much for him. Autopsy failed to find an obvious cause of death. We won't get the opportunity to collect and store sperm from Richard Henry. Fortunately he has three surviving chicks and we also have some of his somatic cells in storage just in case someone figures out how to clone a bird.

# NOTES ON THE SOURCES

The facts of this book were gathered from literature, films, personal experience, and dialogues with those involved. Many of these sources are cited in the text, or in the bibliography, or both. Those quotations not specified as to publication come either from interviews with me, or from historical references and documentaries, a few of which deserve the following special mention.

The history of New Zealand's predator invasion was thoroughly detailed by Carolyn King in her book, *The Immigrant Killers*, on which I heavily relied. And the amended legend of Tibbles was largely informed by Ross Galbreath and Derek Brown in their paper "The tale of the lighthouse-keeper's cat."

Richard Henry's heroic but unheralded life was admirably covered by the authors John and Susanne Hill in their biography, *Richard Henry of Resolution Island*. Don Merton, also a keen student and admirer of Henry, openly shared with me his own appreciations for the great naturalist's uncanny grasp of the indefinable kakapo, and his gallant but tragic history in saving it.

Certain island campaigns were captured, to my great benefit, in films by Roy Hunt (*Battle for Breaksea Island*), Peter and Judy Morrin (*The Battle for Campbell Island*), Scott Mouat (*The Unnatural History of the Kakapo*), and Kevin White (*Restoring Balance: Removing the Black Rat from Anacapa Island.*)

The life and work of Don Merton was gathered in large part

from Alison Ballance's book, *Don Merton: The Man Who Saved the Black Robin,* from David Butler's *Quest for the Kakapo,* and from Merton himself. Additional details were granted by Merton's friends and colleagues.

# BIBLIOGRAPHY

Adams, Douglas, and Mark Carwardine. 1990. *Last Chance to See.* New York: Harmony Books.

Aguirre-Muñoz, Alfonso, Donald A. Croll, C. Josh Donlan, R. William Henry III, Miguel Angel Hermosillo, Gregg R. Howald, Bradford S. Keitt et al. 2008. High-impact conservation: Invasive mammal eradications from the islands of western México. *Ambio* 37 (2): 101–07.

Algar, D. A., A. A. Burbidge, and G. J. Angus. 2002. Cat eradication on Hermite Island, Montebello Islands, western Australia. In *Turning the Tide: The Eradication of Invasive Species,* ed. C. R. Veitch and M. N. Clout, 14–18. Gland, Switzerland and Cambridge, UK: IUCN SSC Invasive Species Specialist Group.

Algar, David, G. John Angus, Rob I. Brazell, Christina Gilbert, and David J. Tonkin. 2003. Feral cats in paradise: Focus on cocos. *Atoll Research Bulletin* 505, 12pp.

Allendorf, Fred W., and Laura L. Lindquist. 2003. Introduction: Population biology, evolution, and control of invasive species. *Conservation Biology* 17 (1): 24–30.

Anderson, Atholl. 2002. Faunal collapse, landscape change and settlement history in Remote Oceania. *World Archaeology* 33 (3): 375–90.

———. 2009. The rat and the octopus: Initial human colonization and the prehistoric introduction of domestic animals in Remote Oceania. *Biological Invasions* 11:1503–19.

Anderson, Paul K. 1995. Competition, predation, and the evolution and extinction of Steller's sea cow, *Hydrodamalis gigas*. *Marine Mammal Science* 11 (3): 391–94.

Angel, Andrea, Ross M. Wanless, and John Cooper. 2009. Review of impacts of the introduced house mouse on islands in the Southern Ocean: Are mice equivalent to rats? *Biological Invasions* 11 (7): 1743–54.

Anon. 2008. *Macquarie Island Pest Eradication Plan: Part H: Project Plan.* Australian Government Department of Environment, Parks, Heritage and the Arts.

———. 2010. *Countdown to Eradication, Macquarie Island Pest Eradication Project Newsletter* 5. Australian Government Department of Environment, Parks, Heritage and the Arts.

Arnaud, Gustavo, Antonio Rodriguez, Alfredo Ortega-Rubio, and Sergio Alvarez-Cardenas. 1993. Predation by cats on the unique endemic lizard of Socorro Island (*Urosaurus auriculatus*), Revillagigedo, Mexico. *Ohio Journal of Science* 93 (4): 101–04.

Asher, Dave, and Dave McCarlie, producers. 2007. *The Titi Islands: A Paradise Restored.* South Coast Productions.

Association for the Study of Animal Behaviour. 2003. Guidelines for the treatment of animals in behavioural research and teaching. *Animal Behaviour* 65:249–55.

Athens, J. Stephen. 2009. *Rattus exulans* and the catastrophic disappearance of Hawai'i's native lowland forest. *Biological Invasions* 11 (7): 1489–1501.

Athens, J. Stephen, Michael W. Kaschko, and Helen F. James. 1991. Prehistoric bird hunters: High altitude resource exploitation on Hawai'i Island. *Bishop Museum Occasional Papers* 31:63–84.

Athens, J. Stephen, H. David Tuggle, Jerome V. Ward, and David J. Welch. 2002. Avifaunal extinctions, vegetation change, and Polynesian impacts in prehistoric Hawai'i. *Archaeology in Oceania* 37:57–78.

Atkinson, I. A. E. 1978. Evidence for the effects of rodents on the vertebrate wildlife of New Zealand Island. In *The Ecology and Control of Rodents in New Zealand Nature Reserves*, ed. P. R. Dingwall, I. A. E. Atkinson, and C. Hay, 33–40. Wellington: Department of Lands and Survey.

Atkinson, Ian A. E. 2001. Introduced mammals and models for restoration. *Biological Conservation* 99:81–96.

Aviss, M. 1997. Post-eradication recovery of Chetwode Island: An update. *Ecological Management* 5:75–77.

Bailey, C. I., and C. T. Eason. 2000. *Anticoagulant Resistance in Rodents.* Wellington: New Zealand Department of Conservation, Conservation Advisory Science Notes 297.

Bailey, Edgar P. 1993. *Introduction of Foxes to Alaskan Islands: History, Effects on Avifauna, and Eradication.* Washington, D.C.: U.S. Department of Interior, Fish and Wildlife Service, Resource Publication, no. 193.

Ballance, Alison. 2007. *Don Merton: The Man Who Saved the Black Robin.* Auckland: Reed Publishing.

Barnes, S. S., E. Matisoo-Smith, and T. L. Hunt. 2006. Ancient DNA of the Pacific rat (*Rattus exulans*) from Rapa Nui (Easter Island). *Journal of Archaeological Science* 33:1536–40.

Baskin, Yvonne. 2003. *A Plague of Rats and Rubbervines: The Growing Threat of Invasive Species.* Washington, D.C.: Island Press.

Beaglehole, Timothy H. 1974. *The Life of Captain James Cook.* London: A. & C. Black Ltd.

Beckerman, A. P., M. Boots, and K. J. Gaston. 2007. Urban bird declines and the fear of cats. *Animal Conservation* 10 (3): 320–25.

Bekoff, Marc. 2002. The importance of ethics in conservation biology: Let's be ethicists not ostriches. *Endangered Species Update* 19 (2): 23–26.

———. 2006. Animal passions and beastly virtues: Cognitive ethology as the unifying science for understanding the subjective, emotional, empathic, and moral lives of animals. *Human Ecology Review* 13 (1): 34–59.

———. 2008. Increasing our compassion footprint: The animals' manifesto. *Zygon* 43 (4): 771–82.

Bekoff, Marc, and Sarah Bexell. 2010. Ignoring nature: Why we do it, the dire consequences, and the need for a paradigm shift to save animals, habitats, and ourselves. *Human Ecology Review* 17 (1): 70–74.

Bell, Brian D. 1978. The Big South Cape islands rat irruption. In *The Ecology and Control of Rodents in New Zealand Nature Reserves*, ed. P. R. Dingwall, I. A. E. Atkinson, and C. Hay, 33–40. Wellington: Department of Lands and Survey.

———. 2002. The eradication of alien mammals from five offshore islands, Mauritius, Indian Ocean. In *Turning the Tide: The Eradication of Invasive Species*, ed. C. R. Veitch and M. N. Clout, 40–45. Gland, Switzerland and Cambridge, UK: IUCN SSC Invasive Species Specialist Group..

Bent, Arthur Cleveland. 1946. *Life Histories of North American Diving Birds*. Binghampton, NY: Dodd, Mead and Company.

Bergstrom, Dana M., Arko Lucieer, Kate Kiefer, Jane Wasley, Lee Belbin, Tore K. Pedersen, and Steven L. Chown. 2009. Indirect effects of invasive species removal devastate World Heritage Island. *Journal of Applied Ecology* 46:73–81.

Bilger, Burkhard. 2009. Swamp things. *New Yorker*, April 20, 80–89.

Black, L.T. 1984. *Atka: An Ethnohistory of the Western Aleutians*. Kingston, Ontario: Limestone Press.

Blackburn, T. M., O. L. Petchey, P. Cassey, and K. J. Gaston. 2005. Functional diversity of mammalian predators and extinction in island birds. *Ecology* 86 (11):2916–23.

Blackburn, T. M., N. Pettorelli, T. Katzner, M. E. Gompper, K. Mock, T. W. J. Garne, R. Altwegg, S. Redpath, and I. J. Gordon. 2010. Dying for conservation: Eradicating invasive alien species in the face of opposition. *Animal Conservation* 13:227–28.

Blackburn, Tim M., Phillip Cassey, Richard P. Duncan, Karl L. Evans, and Kevin J. Gaston. 2004. Avian extinction and mammalian introductions on oceanic islands. *Science* 305 (5692): 1955–58.

Blackburn, Tim M., Phillip Cassey, Richard P. Duncan, Karl L. Evans, and Kevin J. Gaston. 2008. Threats to avifauna on oceanic islands revisited. *Conservation Biology* 22 (2): 492–94.

Booth, William. 2003. On California islets, a clear case of rat and wrong? *Washington Post*, January 5.

Boyer, Alison G. 2008. Extinction patterns in the avifauna of the Hawaiian islands. *Diversity and Distributions* 14 (3): 509–17.

Bremner, A. G., C. F. Butcher, and G. B. Patterson. 1984. The density of indigenous invertebrates on three islands in Breaksea Sound, Fiordland, in relation to the distribution of introduced mammals. *Journal of the Royal Society of New Zealand* 14:379–86.

Brook, F. J. 1999. Changes in the landsnail fauna of Lady Alice Island, northeastern New Zealand. *Journal of the Royal Society of New Zealand* 29:135–57.

———. 2000. Prehistoric predation of the landsnail *Placostylus ambagiosus* Suter (Stylommatophora: Bulimidae) and evidence for the timing of establishment of rats in northernmost New Zealand. *Journal of the Royal Society of New Zealand* 30:227–41.

Brown, James H., and Dov F. Sax. 2004. An essay on some topics concerning invasive species. *Austral Ecology* 29:530–36.

———. 2005. Biological invasions and scientific objectivity: Reply to Cassey et al. (2005). *Austral Ecology* 30:481–83.

Brown, K. P. 1997. Predation at nests of two New Zealand endemic passerines; implications for bird community restoration. *Pacific Conservation Biology* 3:91–98.

Bryant, David M. 2006. Energetics of free-living kakapo (*Strigops habroptilus*). *Notornis* 53 (1): 126–37.

Buckelew, S., G. Howald, S. MacLean, V. Byrd, L. Daniel, S. Ebbert, and W. Meeks. 2009. *Rat Island Habitat Restoration*

*Project: Operational Report*. Report to USFWS. Island Conservation, Santa Cruz, CA.

Buckelew, Stacey, Gregg Howald, Don Croll, Steve MacLean, Vernon Byrd, Steve Ebbert, Jen Curl et al. *Invasive Rat Eradication on Rat Island, Aleutian Islands, Alaska: Biological Monitoring and Operational Assessment*. Unpublished report, 65pp.

Buckle, A. P., and M. G. P. Fenn. 1992. Rodent control in the conservation of endangered species. In *Proceedings of the Fifteenth Vertebrate Pest Conference*, ed. J. E. Borrecco and R. E. Marsh, 36–41. Davis: University of California.

Buckle, A. P., C. V. Prescott, and K. J. Ward. 1994. Resistance to the first and second generation anticoagulant rodenticides—a new perspective. In *Proceedings of the Sixteenth Vertebrate Pest Conference*, ed. W. S. Halverson and A. C. Crabb, 138–44. Davis: University of California.

Buller, Walter L. 1878. Further notes on the ornithology of New Zealand. *Transactions of the Royal Society of New Zealand* 10:201–09.

———. 1895. Stephen Island Wren. *Ibis* 7 (1): 236–37.

———. 1905. *Supplement to the 'Birds of New Zealand.'* 2 vols. Published by the author, London.

Burdick, Alan. 2005. *Out of Eden: An Odyssey of Ecological Invasion*. New York: Farrar, Straus and Giroux.

Burghardt, Gordon M., and Harold A. Herzog Jr. 1980. Beyond conspecifics: Is Brer Rabbit our brother? *Bioscience* 30 (11): 763–68.

Burney, David, Helen F. James, Lida Pigott Burney, Storrs L. Olson, William Kikuchi, Warren L. Wagner, Mara Burney et al. 2001. Fossil evidence for a diverse biota from Kaua'i and its transformation since human arrival. *Ecological Monographs* 71 (4): 615–41.

Burney, David A., and Timothy F. Flannery. 2005. Fifty millennia of catastrophic extinctions after human contact. *Trends in Ecology and Evolution* 20 (7): 395–401.

Butler, D. J. 2006. The habitat, food and feeding ecology of kakapo in Fiordland: A synopsis from the unpublished MSc thesis of Richard Gray. *Notornis* 53 (1): 55–79.

Butler, David J. 1989. *Quest for the Kakapo.* Auckland: Heinemann Reed.

Byrd, G. V. 1987. *Wildlife Survey of Walrus Island, Pribilof Islands, Alaska.* Unpublished report, U.S. Fish and Wildlife Service, Adak, Alaska, 7pp.

Campbell, D. J. 1978. The effects of rats on vegetation. In *The Ecology and Control of Rodents in New Zealand Nature Reserves*, ed. P. R. Dingwall, I. A. E. Atkinson, and C. Hay, 99–120. Wellington: Department of Lands and Survey.

Campbell, D. J., and I. A. E. Atkinson. 1999. Effects of kiore (*Rattus exulans*) on recruitment of indigenous coastal trees on northern offshore islands of New Zealand. *Journal of the Royal Society of New Zealand* 29:265–90.

———. 2002. Depression of tree recruitment by the Pacific rat (*Rattus exulans* Peale) on New Zealand's northern offshore islands. *Biological Conservation* 107:19–35.

Campbell, Karl, and C. Josh Donlan. 2005. Feral goat eradications on islands. *Conservation Biology* 19 (5): 1362–74.

Campbell, Karl, C. Josh Donlan, F. Cruz, and V. Carrion. 2004. Eradication of feral goats (*Capra hircus*) from Pinta Island, Galápagos, Ecuador. *Oryx* 38:328–33.

Campbell, Mike. 2008. Biologists hope to kill all of Rat Island's rats. *Anchorage Daily News*, September 21.

Canby, Thomas Y. 1977. The rat: Lapdog of the devil. *National Geographic*, July, 60–87.

Carey, Benedict. 2006. Message from mouse to mouse: I feel your pain. *New York Times*, July 4. http://www.nytimes.com/2006/07/04/health/04empa.html (accessed June 23, 2010).

Carrion, Victor, C. Josh Donlan, Karl Campbell, Christian Lavoie, and Felipe Cruz. 2007. Feral donkey (*Equus asinus*) eradication in the Galápagos. *Biodiversity and Conservation* 16:437–45.

Case, T. J., M. L. Cody, and E. Ezcurra. 2002. *A New Island Biogeography of the Sea of Cortes.* New York: Oxford University Press.

Caut, Stéphane, Elena Angulo, and Franck Courchamp. 2008. Dietary shift of an invasive predator: Rats, seabirds and sea turtles. *Journal of Applied Ecology* 45:428–37.

Caut, Stéphane, Jorge G. Casanovas, Emilio Virgos, Jorge Lozano, Gary W. Witmer, and Franck Courchamp. 2007. Rats dying for mice: Modelling the competitor release effect. *Austral Ecology* 32:858–68.

Chadwick, Douglas H. 2006. Life in the desert: Songs of the Sonoran. *National Geographic*, September. http://science.nationalgeographic.com/science/earth/habitats/sonoran-desert.html?nav =FEATURES (accessed Nov. 17, 2010).

Chew, Matthew K., and Manfred D. Laubichler. 2003. Natural enemies—metaphor or misconception? *Science* 301:52–53.

Chinery, Michael. 1990. *Predators: Killers of the Wild.* London: Bedford Editions.

Clapperton, B. Kay. 2006: A review of the current knowledge of rodent behaviour in relation to control devices. *Science for Conservation* 263, 55pp.

Clavero, Miguel, and Emili García-Berthou. 2005. Invasive species are a leading cause of animal extinctions. *Trends in Ecology and Evolution* 20 (3): 110.

Clayton, Richard I., Deborah J. Wilson, Katharine J. M. Dickinson, and Carol J. West. 2008. Response of seedling communities to mammalian pest eradication on Ulva Island, Rakiura National Park, New Zealand. *New Zealand Journal of Ecology* 32 (1): 103–07.

Clout, M. N. 2003. Biodiversity loss caused by invasive alien vertebrates. *Zeitschrift für Jagdwissenschaft* 48 (supp. 1): 51–58.

Clout, M. N., and C. R. Veitch. 2002. Turning the tide of biological invasion: The potential for eradicating invasive species. In

*Turning the Tide: The Eradication of Invasive Species*, ed. C. R. Veitch and M. N. Clout, 1–3. Gland, Switzerland and Cambridge, UK: IUCN SSC Invasive Species Specialist Group.

Clout, Mick N. 2006. A celebration of kapapo: Progress in the conservation of an enigmatic parrot. *Notornis* 53 (1): 1–2.

Clout, Mick N., and James C. Russell. 2007. The invasion ecology of mammals: A global perspective. *Wildlife Research* 35:180–84.

Cockrem, John F. 2006. The timing of breeding in the kakapo (*Strigops habroptilus*). *Notornis* 53 (1): 153–59.

Coghlan, Andy. 2009. In search of the perfect mouse trap. *New Scientist*, July. http://www.newscientist.com/article/dn17519-.html? (accessed June 21, 2010).

Colautti, Robert I., and Hugh J. MacIsaac. 2004. A neutral terminology to define 'invasive' species. *Diversity and Distributions* 10:135–41.

Collerson, Kenneth D., and Marshall I. Weisler. 2007. Stone adze compositions and the extent of ancient Polynesian voyaging and trade. *Science* 317:1907–11.

Connor, Steve. 2002. Ecologists hail victory over Galapagos invader. *Independent* (UK), June 3. http://www.independent.co.uk/environment/ecologists-hail-victory-over-galapagos-invader-644292.html (accessed June 2, 2009).

Cottam, Yvette, Don V. Merton, and Wouter Hendriks. 2006. Nutrient composition of the diet of parent-raised kakapo nestlings. *Notornis* 53 (1): 90–99.

Coues, Elliott. 1868. A monograph of the Alcidae. *Proceedings of the Academy of Natural Science of Philadelphia* 2 (20): 2–81.

Courchamp, F., and S. Caut. 2005. Use of biological invasions and their control to study the dynamics of interacting populations. In *Conceptual Ecology and Invasions Biology*, ed. M. W. Cadotte, S. M. McMahon, and T. Fukami, 253–79. Great Britain: Springer.

Courchamp, Franck, Jean-Louis Chapuis, and Michel Pascal. 2003. Mammal invaders on islands: Impact, control and control impact. *Biological Reviews* 78:347–83.

Courchamp, Franck, Michel Langlais, and George Sugihara. 1999. Cats protecting birds: Modelling the mesopredator release effect. *Journal of Animal Ecology* 68:282–92.

———. 2000. Rabbits killing birds: Modelling the hyperpredation process. *Journal of Animal Ecology* 69:154–64.

Courchamp, Franck, and George Sugihara. 1999. Modeling the biological control of an alien predator to protect island species from extinction. *Ecological Applications* 9 (1): 112–23.

Cowan, P. E. 1992. The eradication of introduced Australian brushtail possums, *Trichosurus vulpecula*, from Kapiti Island, a New Zealand nature reserve. *Biological Conservation* 61:217–26.

Cree, A., C. H. Daugherty, and J. M. Hay. 1995. Reproduction of a rare reptile, the tuatara *Sphenodon punctatus*, on rat-free and rat-inhabited islands in New Zealand. *Conservation Biology* 9:373–83.

Croll, D. A., J. L. Maron, J. A. Estes, E. M. Danner, and G. V. Byrd. 2005. Introduced predators transform subarctic islands from grassland to tundra. *Science* 307:1959–61.

Cromarty, P. L., K. G. Broome, A. Cox, R. A. Empson, W. M. Hutchinson, and I. McFadden. 2002. Eradication planning for invasive alien animal species on islands—the approach developed by the New Zealand Department of Conservation. In *Turning the Tide: The Eradication of Invasive Species*, ed. C. R. Veitch and M. N. Clout, 85–91. Gland, Switzerland and Cambridge, UK: IUCN SSC Invasive Species Specialist Group.

Crook, I. G. 1973. The tuatara, *Sphenodon punctatus* Gray, on islands with and without populations of the Polynesian rat, *Rattus exulans* (Peale). *Proceedings of the New Zealand Ecological Society* 20:115–20.

Cruz, Felipe, C. Josh Donlan, Karl Campbell, and Victor Carrion. 2005. Conservation action in the Galápagos: Feral pig (*Sus scrofa*) eradication from Santiago Island. *Biological Conservation* 121:473–78.

Curwood, Steve. 2009. Taking the "Rat" out of Rat Island. *Living on Earth*, July 17. http://www.loe.org/shows/segments.htm?programID=09-P13-00029&segmentID=5 (accessed Nov. 17, 2010).

Cuthbert, Richard J., and Erica S. Sommer. 2004. Population size and trends of four globally threatened seabirds at Gough Island, South Atlantic Ocean. *Marine Ornithology* 32:97–103.

Daniel, M. J., and G. R. Williams. 1984. A survey of the distribution, seasonal activity and roost sites of New Zealand bats. *New Zealand Journal of Ecology* 7:9–25.

Darwin, Charles. 1958. *The Origin of the Species By Means of Natural Selection of the Preservation of Favoured Races in the Struggle for Life.* New York: Mentor.

Daugherty, Charles, and David Town. 1991. The cat's breakfast. *New Zealand Science Monthly* (April): 13–14.

Dawkins, Marian Stamp. 2008. The science of animal suffering. *Ethology* 114:937–45.

Decety, Jean. 2010. To what extent is the experience of empathy mediated by shared neural circuits? *Emotion Review* 2 (3): 204–07.

DeWaal, Frans B. M. 2008. Putting the altruism back into altruism: The evolution of empathy. *Annual Review of Psychology* 59:279–300.

Diamond, Jared. 1985. Rats as agents of extermination. *Nature* 318: 602–03.

———. 1995. Easter's end. *Discover* 9:62–69.

———. 2005. *Collapse: How Societies Choose to Fail or Succeed.* New York: Viking.

———. 2007. Easter Island revisited. *Science* 317:1692–94.

Diamond, Jared M., and C. Richard Veitch. 1981. Extinctions and introductions in the New Zealand avifauna: Cause and effect? *Science* 211:499–501.

Didham, Raphael K., Robert M. Ewers, and Neil J. Gemmell. 2005. Comment on "Avian extinction and mammalian introductions on oceanic islands." *Science* 307:1412.

Dilks, Peter, and David Towns. 2002. *Developing Tools to Detect and*

*Respond to Rodent Invasions of Islands: Workshop Report and Recommendations*. Wellington: New Zealand Department of Conservation, DOC Science Internal Series 59.

Dingwall, P. R., I. A. E. Atkinson, and C. Hay, eds. 1978. *The Ecology and Control of Rodents in New Zealand Nature Reserves*. Information Series, 4. Wellington: Department of Lands and Survey.

Donlan, C. J., and C. Wilcox. Bringing invasive mammal eradication into the conservation limelight. *Wildlife Research* (in press).

Donlan, C. Josh. 2008. Rewilding the islands. In *State of the Wild 2008–2009: A Global Portrait of Wildlife, Wildlands, and Oceans*, ed. Eva Fearn, 226–32. Washington, D.C.: Island Press.

Donlan, C. Josh, Karl Campbell, Wilson Cabrera, Christian Lavoie, Victor Carrion, and Felipe Cruz. 2007. Recovery of the Galápagos rail (*Laterallus spilonotus*) following the removal of invasive mammals. *Biological Conservation* 138:520–24.

Donlan, C. Josh, Donald A. Croll, and Bernie R. Tershy. 2003. Islands, exotic herbivores, and invasive plants: Their roles in coastal California restoration. *Restoration Ecology* 11 (4): 524–30.

Donlan, C. Josh, Gregg R. Howald, Bernie R. Tershy, and Donald A. Croll. 2003. Evaluating alternative rodenticides for island conservation: Roof rat eradication from the San Jorge Islands, Mexico. *Biological Conservation* 114:29–34.

Donlan, C. Josh, Bernie R. Tershy, Karl Campbell, and Felipe Cruz. 2003. Research for requiems: The need for more collaborative action in eradication of invasive species. *Conservation Biology* 17 (6): 1850–51.

Donlan, C. Josh, Bernie R. Tershy, and Donald A. Croll. 2002. Islands and introduced herbivores: Conservation action as ecosystem experimentation. *Journal of Applied Ecology* 39:235–46.

Donlan, C. Josh, Bernie R. Tershy, Brad S. Keitt, Bill Wood, José Ángel Sánchez, Anna Weinstein, Donald A. Croll, Miguel Ángel Hermosillo, and José Luis Aguilar. 2000.Island conservation action in northwest México. In *Proceedings of the Fifth*

*California Islands Symposium*, ed. D. R. Browne, K. L. Mitchell, and H. W. Chaney, 330–38. Santa Barbara: Santa Barbara Museum of Natural History.

Donlan, C. Josh, and Chris Wilcox. 2008. Diversity, invasive species and extinctions in insular ecosystems. *Journal of Applied Ecology* 45:1114–23.

———. 2008. Integrating invasive mammal eradications and biodiversity offsets for fisheries bycatch: Conservation opportunities and challenges for seabirds and sea turtles. *Biological Invasions* 10 (7): 1053–60.

Drake, Donald R., and Terry L. Hunt. 2009. Invasive rodents on islands: Integrating historical and contemporary ecology. *Biological Invasions* 11:1483–87.

Drever, Mark Christopher. 1997. Ecology and eradication of Norway rats on Langara Island, Queen Charlotte Islands. Master's thesis, Simon Fraser University.

———. 2000. Birds back, rats razed. *SpruceRoots Magazine*, February. http://www.spruceroots.org/February%202000/BirdsBack.html (accessed Sept. 12, 2008).

Driscoll, Carlos A., Juliet Clutton-Brock, Andrew C. Kitchener, and Stephen J. O'Brien. 2009. The evolution of house cats. *Scientific American*, June 10. http://www.scientificamerican.com/article.cfm?id=the-taming-of-the-cat (accessed Oct. 1, 2010).

Driscoll, Carlos A., Marilyn Menotti-Raymond, Alfred L. Roca, Karsten Hupe, Warren E. Johnson, Eli Geffen, Eric H. Harley et al. 2007. The Near Eastern origin of cat domestication. *Science* 317:519–23.

Duncan, Richard P., Tim M. Blackburn, and Trevor H. Worthy. 2002. Prehistoric bird extinctions and human hunting. *Proceedings of the Royal Society of London* B 269: 517–21.

Eason, C. T., E. C. Murphy, G. R. Wright, and E. B. Spurr. 2002. Assessment of risks of brodifacoum to non-target birds and mammals in New Zealand. *Ecotoxicology* 11:35–48.

Eason, Daryl, and Ron J. Moorhouse. 2006. Hand-rearing kakapo (*Strigops habroptilus*), 1997–2005. *Notornis* 53 (1): 116–25.

Eason, Daryl K., Graeme P. Elliott, Don V. Merton, Paul W. Jansen, Grant A. Harper, and Ron J. Moorhouse. 2006. Breeding biology of kakapo (*Strigops habroptilus*) on offshore island sanctuaries, 1990–2002. *Notornis* 53 (1): 27–36.

Ebbert, Steve E., and G. Vernon Byrd. 2002. Eradications of invasive mammals to restore natural biological diversity on Alaska Maritime National Wildlife Refuge. In *Turning the Tide: The Eradication of Invasive Species*, ed. C. R. Veitch and M. N. Clout, 102–09. Gland, Switzerland and Cambridge, UK: IUCN SSC Invasive Species Specialist Group.

Ebbert, Steve, and Kathy Huntington. 2010. Anticoagulant residual concentration and poisoning in birds following a large-scale aerial application of 25-ppm brodifacoum bait for rat eradication on Rat Island, Alaska. Paper presented at Twenty-fourth Vertebrate Pest Conference, Sacramento, California, February 22–25.

Elliott, Graeme P. 2006. A simulation of the future of kakapo. *Notornis* 53 (1): 164–72.

Elliott, Graeme P., Daryl K. Eason, Paul W. Jansen, Don V. Merton, Grant A. Harper, and Ron J. Moorhouse. 2006. Productivity of kakapo (*Strigops habroptilus*) on offshore island refuges. *Notornis* 53 (1): 138–42.

Elton, Charles. 2000. *The Ecology of Invasions by Animals and Plants*. Chicago: Chicago University Press.

Elwood, Robert W. 1991. Ethical implications of studies on infanticide and maternal aggression in rodents. *Animal Behavior* 42:841–49.

Farrimond, Melissa, Mick N. Clout, and Graeme P. Elliott. 2006. Home range size of kakapo (*Strigops habroptilus*) on Codfish Island. *Notornis* 53 (1): 150–52.

Farrimond, Melissa, Graeme P. Elliott, and Mick N. Clout. 2006. Growth and fledging of kakapo. *Notornis* 53 (1): 112–15.

Feare, C. 1999. Ants take over from rats on Bird Island, Seychelles. *Bird Conservation International* 9:95–96.

Fitzgerald, B. M., B. J. Karl, and C. R. Veitch. 1991. The diet of feral cats (*Felis catus*) on Raoul Island, Kermadec Group. *New Zealand Journal of Ecology* 15 (2): 123–29.

Fleming, C. A. 1969. Rats and moa extinction. *Notornis* 16 (3): 210–11.

Ford, Corey. 1966. *Where the Sea Breaks Its Back*. Anchorage: Alaska Northwest Books.

Frost, Orcutt. 2003. *Bering: The Russian Discovery of America*. New Haven, CT: Yale University Press.

Fukami, Tadashi, David A. Wardle, Peter J. Bellingham, Christa P. H. Mulder, David R. Towns, Gregor W. Yeates, Karen I. Bonner, Melody S. Durrett, Madeline N. Grant-Hoffman, and Wendy M. Williamson. 2006. Above- and below-ground impacts of introduced predators in seabird-dominated island ecosystems. *Ecology Letters* 9 (12): 1299–1307.

Fuller, Errol. 2000. *Extinct Birds*. 2nd ed. Oxford: Oxford University Press.

Funk, Caroline. 2010. *Rats and Birds: Tracking Ecological Change with Evidence from Prehistoric to Historic Aleut Village Midden Test Excavations*, Rat Island, Alaska. Unpublished report, 79pp.

Galbreath, Ross, and Derek Brown. 2004. The tale of the lighthouse-keeper's cat: Discovery and extinction of the Stephens Island wren (*Traversia lyalli*). *Notornis* 51 (4): 193–200.

Garfield, Brian. 1982. *The Thousand-Mile War*. New York: Bantam Books.

Gaston, Anthony J. 2004. *Seabirds: A Natural History*. New Haven, CT: Yale University Press.

Gobster, Paul H. 2005. Invasive species as ecological threat: Is restoration an alternative to fear-based resource management? *Ecological Restoration* 23 (4): 261–70.

Goldstein, Jared A. 2008. Aliens in the garden. *Roger Williams University School of Law Faculty Papers*, 57pp.

Golumbia, Todd E. 1999. Introduced species management in Haida Gwaii (Queen Charlotte Islands). In *Proceedings of a Conference on the Biology and Management of Species and Habitats at Risk*, Kamloops, B.C., February 15–19.

Graham, M. F., and C. R. Veitch. 2002. Changes in bird numbers on Tiritiri Matangi Island, New Zealand, over the period of rat eradication. In *Turning the Tide: The Eradication of Invasive Species*, ed. C. R. Veitch and M. N. Clout, 120–23. Gland, Switzerland and Cambridge, UK: IUCN SSC Invasive Species Specialist Group.

Grant, Gilbert S., Ted N. Pettit, and G. Causey Whittow. 1981. Rat predation on Bonin petrel eggs on Midway Atoll. *Journal of Field Ornithology* 52 (4): 336–38.

Green, C. 2002. Recovery of invertebrate populations on Tiritiri Matangi Island, New Zealand following eradication of Pacific rats (*Rattus exulans*). In *Turning the Tide: The Eradication of Invasive Species*, ed. C. R. Veitch and M. N. Clout, 407. Gland, Switzerland and Cambridge, UK: IUCN SSC Invasive Species Specialist Group.

Griggs, Kim. 2005. Winging it. *NZ Listener* 198 (3389): 23–29.

Grzelewski, Derek. 2002. Going to extremes. *Smithsonian*, October, 90–95.

Gurevitch, Jessica. 2006. Commentary on Simberloff (2006): Meltdowns, snowballs and positive feedbacks. *Ecology Letters* 9 (8): 919–21.

Gurevitch, J., and D. K. Padilla. 2004. Are invasive species a major cause of extinctions? *Trends in Ecology and Evolution* 19:470–74.

Guthrie-Smith, W. H. 1936. *Sorrows and Joys of a New Zealand Naturalist*. Wellington: A. H. & A. W. Reed.

Haami, Bradford. 1994. The *kiore* rat in Aotearoa: A Maori perspective. In *Science of Pacific Island Peoples: Fauna, Flora, Food, and Medicine*, ed. R. J. Morrison, Paul A. Geraghty, and Linda Crowl, 65-76. Suva, Fiji: University of the South Pacific.

Hadler, Malcolm R., and Alan P. Buckle. 1992. Forty five years of anticoagulant rodenticides—past, present and future trends. In *Proceedings of the Fifteenth Vertebrate Pest Conference*, ed. J. E. Borrecco and R. E. Marsh, 149–155. Davis: University of California.

Harper, Grant A., Graeme P. Elliott, Daryl K. Eason, and Ron J. Moorhouse. 2006. What triggers nesting of kakapo (*Strigops habroptilus*)? *Notornis* 53:160–63.

Harper, Grant A., and Joanne Joice. 2006. Agonistic display and social interaction between female kakapo (*Strigops habroptilus*). *Notornis* 53:195–97.

Harris, Donna B., and David W. Macdonald. 2007. Interference competition between introduced black rats and endemic Galápagos rice rats. *Ecology* 88 (9): 2330–44.

Harrison, Craig S. 1992. A conservation agenda for the 1990s: Removal of alien predator from seabird colonies. *Pacific Seabird Group Bulletin* 19 (1): 5.

Harrison, Peter. 1983. *Seabirds: An Identification Guide.* Boston: Houghton Mifflin Company.

Herzog, Harold A., and Lauren L. Golden. 2009. Moral emotions and social activism: The case of animal rights. *Journal of Social Issues* 65 (3): 485–98.

Hill, Susanne, and John Hill. 1987. *Richard Henry of Resolution Island.* Dunedin, New Zealand: John McIndoe.

Hoare, Joanne M., Lynn K. Adams, Leigh S. Bull, and David R. Towns. 2007. Attempting to manage complex predator–prey interactions fails to avert imminent extinction of a threatened New Zealand skink population. *Journal of Wildlife Management* 71 (5): 1576–84.

Holdaway, R. N. 1989. New Zealand's pre-human avifauna and its vulnerability. *New Zealand Journal of Ecology* 12:11–25.

———. 1996. Arrival of rats in New Zealand. *Nature* 384:225–26.

———. 1999. Introduced predators and avifaunal extinctions in New Zealand. In *Extinctions in Near Time*, ed. R. D. E.

MacPhee, 189–238. New York: Academic/Plenum Publishers.

———. 1999. A spatio-temporal model for the invasion of the New Zealand archipelago by the Pacific rat *Rattus exulans*. *Journal of the Royal Society of New Zealand* 29 (2): 91–105.

Horwitz, Tony. 2002. *Blue Latitudes: Boldly Going Where Captain Cook Had Gone Before*. New York: Henry Holt.

Howald, G. R., P. Mineau, J. E. Elliott, and K. M. Cheng. 1999. Brodifacoum poisoning of avian scavengers during rat control on a seabird colony. *Ecotoxicology* 8:431–47.

Howald, Gregg, C. Josh Donlan, Juan Pablo Galván, James C. Russell, John Parkes, Araceli Samaniego, Yiwei Wang et al. Invasive rodent eradication on islands. 2007. *Conservation Biology* 21 (5): 1258–68.

Howald, Gregg R., Kate R. Faulkner, Bernie Tershy, Bradford Keitt, Holly Gellerman, Eileen M. Creel, Matthew Grinnell, Steven T. Ortega, and Donald A. Croll. 2005. Eradication of black rat from Anacapa Island: Biological and social considerations. In *Proceedings of the Sixth California Islands Symposium*, ed. D. K. Garcelon and C. A. Schwemm, 299–312. Arcata, CA: Institute for Wildlife Studies.

Hughes, B. John, Graham R. Martin, and S. James Reynolds. 2008. Cats and seabirds: Effects of feral domestic cat *Felis silvestris catus* eradication on the population of sooty terns *Onychoprion fuscata* on Ascension Island, South Atlantic. *Ibis* 150 (supp. 1): 122–31.

Humane Vertebrate Pest Control Working Group. 2004. A national approach towards humane vertebrate pest control. In *Proceedings of an RSPCA Australia/AWC/VPC Joint Workshop*, August 4–5, Melbourne, Canberra: RSPCA Australia.

Hunt, George L., and Nancy M. Harrison. 1990. Foraging habitat and prey taken by least auklets at King Island, Alaska. *Marine Ecology Progress Series* 65:141–50.

Hunt, George L. Jr., R. L. Pitman, and H. Lee Jones. 1980. Distribution and abundance of seabirds breeding on the California Channel Islands. In *2nd California Islands Multidisciplinary Symposium*, ed. D. M. Power, 443–59.

Hunt, Roy, producer. 1989. *Battle for Breaksea Island*. Natural History NZ.

Hunt, Terry L. 2007. Rethinking Easter Island's ecological catastrophe. *Journal of Archaeological Science* 34:485–502.

Hurley, Timothy. 2002. Cats suspected in sea-bird kill on Maui. *Honolulu Advertiser*, Aug. 9. http://the.honoluluadvertiser.com/article/2002/Aug/09/ln/ln23a.html (accessed Aug. 11, 2009).

Hutchins, Michael. 2007. The limits of compassion. *The Wildlife Professional*, Summer:42–44.

Imber, M., M. Harrison, and J. Harrison. 2000. Interactions between petrels, rats and rabbits on Whale Island, and effects of rat and rabbit eradication. *New Zealand Journal of Ecology* 24:153–60.

Imber, M. J. 1978. The effects of rats on breeding success of petrels. In *The Ecology and Control of Rodents in New Zealand Nature Reserves*, ed. P. R. Dingwall, I. A. E. Atkinson, and C. Hay, 67–71. Wellington: Department of Lands and Survey.

Iwaniuk, Andrew N., Storrs L. Olson, and Helen F. James. 2009. Extraordinary cranial specialization in a new genus of extinct duck (Aves: Anseriformes) from Kauai, Hawaiian Islands. *Zootaxa* 2296:47–67.

Jackson, Philip L., Pierre Rainville, and Jean Decety. 2006. To what extent do we share the pain of others? Insight from the neural bases of pain empathy. *Pain* 125:5–9.

Jackson, William B. 1967. Rats, bombs, and paradise—the story at Eniwetok. In *Proceedings of the Third Vertebrate Pest Conference*, 45–46. Davis: University of California.

———. 1980. Rats: friends or foes? *Journal of Popular Culture* 14:27–32.

Jackson, William B., and Dale E. Kaukeinen. 1972. The problem of anticoagulant rodenticide resistance in the United States. In

*Proceedings of the Fifth Vertebrate Pest Conference,* 142–48.Davis: University of California.

James, Helen F., and Storrs L. Olson. 1983. Flightless birds. *Natural History* 92 (9): 30–40.

Jansen, Paul W. 2006. Kakapo recovery: The basis of decision-making. *Notornis* 53:184–90.

Jones, Holly P., Bernie R. Tershy, Erika S. Zavaleta, Donald A. Croll, Bradford S. Keitt, Myra E. Finkelstein, and Gregg R. Howald. 2008. Severity of the effects of invasive rats on seabirds: A global review. *Conservation Biology* 22 (1): 16–26.

Jones, Ian L. 1993. Least auklet (*Aethia pusilla*). In *The Birds of North America*, no. 69, ed. A. Poole and F. Gill. Ithaca: Cornell Lab of Ornithology

Jones, Ian L., and Montgomerie R. 1991. Mating and remating in least auklets (*Aethia pusilla*) relative to ornamental traits. *Behavioral Ecology* 2:249–57.

———. 1992. Least auklet ornaments: Do they function as quality indicators? *Behavioural Ecology & Sociobiology* 30:43–52.

Jost, Christian H., and Serge Andrefouët. 2006. Long term natural and human perturbations and current status of Clipperton Atoll, a remote island of the Eastern Pacific. *Pacific Conservation Biology* 12 (3): 207–18.

Jouventin, Pierre, Joël Bried, and Thierry Micol. 2003. Insular bird populations can be saved from rats: A long-term experimental study of white-chinned petrels *Procellaria aequinoctialis* on Ile de la Possession (Crozet archipelago). *Polar Biology* 26:371–78.

Kapa, D. 2003. The eradication of kiore and the fulfilment of kaitiakitanga obligations. *Auckland University Law Review* 9:1326–52.

Karels, Tim J., F. Stephen Dobson, Heather S. Trevino, and Amy L. Skibiel. 2008. The biogeography of avian extinctions on oceanic islands. *Journal of Biogeography* 35:1106–11.

Kaukeinen, Dale E., and Michael Rampaud. 1986. A review of brodifacoum efficacy in the U.S. and worldwide. In *Proceedings of the Twelfth Vertebrate Pest Conference*, ed. T. P. Salmon, 16–50. Davis: University of California.

Keitt, Bradford S. 2005. Status of Xantus's murrelet and its nesting habitat in Baja California, Mexico. *Marine Ornithology* 33:105–14.

Keitt, Bradford S., Chris Wilcox, Bernie R. Tershy, Donald A. Croll, and C. Josh Donlan. 2002. The effect of feral cats on the population viability of black-vented shearwaters (*Puffinus opisthomelas*) on Natividad Island, Mexico. *Animal Conservation* 5:217–23.

Kepler, Cameron B. 1967. Polynesian rat predation on nesting Laysan albatrosses and other Pacific seabirds. *Auk* 84:426–30.

Kettman, Matt. 2003. Death for life on Anacapa Island. *Santa Barbara Independent*, April 29.

King, C. M. 1983. The relationship between beech (*Nothofagus* sp.) seedfall and populations of mice (*Mus musculus*), and the demographic and dietary responses of stoats (*Mustela erminea*) in three New Zealand forests. *Journal of Animal Ecology* 52:414–66.

King, Carolyn. 1985. *The Immigrant Killers*. Auckland: Oxford University Press.

Kirk, R. L. 1987. Population scenarios: Prehistory in the Pacific Islands. *Science* 235:1683–84.

Kirkwood, J. K., A. W. Sainsbury, and P. M. Bennett. 1994. The welfare of free-living wild animals: Methods of assessment. *Animal Welfare* 3:257–73.

Knowlton, Jessie L., C. Josh Donlan, Gary W. Roemer, Araceli Samaniego-Herrera, Bradford S. Keitt, Bill Wood, Alfonso Aguirre-Muñoz, Kate R. Faulkner, and Bernie R. Tershy. 2007. Eradication of non-native mammals and the status of insular mammals on the California Channel Islands, USA, and Pacific Baja California Peninsula Islands, Mexico. *Southwestern Naturalist* 52 (4): 528–40.

Koch, Paul L., and Anthony D. Barnosky. 2006. Late Quaternary extinctions: State of the debate. *Annual Review of Ecology, Evolution, and Systematics* 37:215–50.

Kurle, Carolyn M., Donald A. Croll, and Bernie R. Tershy. 2008. Introduced rats indirectly change marine rocky intertidal communities from algae- to invertebrate-dominated. *Proceedings of the National Academy of Sciences* 105 (10): 3800–3804.

Langford, Dale J., Sara E. Crager, Zarrar Shehzad, Shad B. Smith, Susana G. Sotocinal, Jeremy S. Levenstadt, Mona Lisa Chanda, Daniel J. Levitin, and Jeffrey S. Mogil. 2006. Social modulation of pain as evidence for empathy in mice. *Science* 312:1967–70.

Larson, Brendon M. H. 2005. The war of the roses: Demilitarizing invasion biology. *Frontiers in Ecology and the Environment* 3 (9): 495–500.

———. 2007. Who's invading what? Systems thinking about invasive species. *Canadian Journal of Plant Science* 87:993–99.

Lavoie, C., C. J. Donlan, K. Campbell, F. Cruz, and V. Carrion. 2007. Geographic tools for facilitating the management of insular non-native mammal eradication programs. *Biological Invasions* 9:139–48.

Lawton, J. H. 1997. The science and non-science of conservation biology. *Oikos* 79:3–5.

LeCorre, Matthieu. 2008. Cats, rats and seabirds. *Nature* 451:134–35.

Lessard, Robert B., Steven J. D. Martell, Carl J. Walters, Timothy E. Essington, and James F. Kitchell. 2005. Should ecosystem management involve active control of species abundances? *Ecology and Society* 10 (2): 1. http://www.ecologyandsociety.org/vol10/iss2/art1 (accessed June 3, 2009).

Line, Les. 1995. Gone the way of the dinosaurs: A scientist unearths evidence of a recent mass extinction in which 2,000 species of rails and other birds disappeared in the South Pacific. *International Wildlife*, July/August:16–21.

Littin, K. E., and D. J. Mellor. 2005. Strategic animal welfare issues: Ethical and animal welfare issues arising from the killing of wildlife for disease control and environmental reasons. *Revue Scientifique et Technique–Office International des Epizooties* 24 (2): 767–82.

Littin, K. E., C. E. O'Connor, and C. T. Eason. 2000. Comparative effects of brodifacoum on rats and possums. *New Zealand Plant Protection* 53:310–15.

Little, Jane Braxton. 2006. Treasure Island. *Audubon*, September:75–78, 93.

Llana, Sara Miller, and Moises Velasquez-Manoff. 2009. Saving the Galapagos means rebuilding nature. *Christian Science Monitor*, April 19. http://www.csmonitor.com/2009/0419/p07s02-wogn .html (accessed June 2, 2009).

Lockwood, Jeffrey, and Alexandre V. Latchininsky. 2008. Confessions of a Hit Man. *Conservation* 9 (3). http://www.conservationmagazine.org/articles/v9n3/confessions-of-a-hit-man (accessed May 3, 2010).

Lovegrove, T. G. 1996. A comparison of the effects of predation by Norway (*Rattus norvegicus*) and Polynesian rats (*R. exulans*) on the saddleback (*Philesturnus carunculatus*). *Notornis* 43:91–112.

Lowe, David J. 2008. Polynesian settlement of New Zealand and the impacts of volcanism on early Maori society: An update. In *Guidebook for Pre-conference North Island Field Trip A1 'Ashes and Issues,'* 142–47, ed. D. J. Lowe. Australian and New Zealand 4th Joint Soils Conference, Massey University. Palmerston North: New Zealand Society of Soil Science.

Mansfield, Bill, and David Towns. 1997. Lesson of the islands. *Restoration and Management Notes* 15 (2): 138–46.

Maron, John L., James A. Estes, Donald A. Croll, Eric M. Danner, Sarah C. Elmendorf, and Stacey L. Buckelew. 2006. An introduced predator transforms Aleutian Island plant communities by disrupting spatial subsidies. *Ecological Monographs* 76 (1): 3–24.

Martin, A. R., S. Poncet, C. Barbraud, E. Foster, P. Fretwell, and P. Rothery. 2009. The white-chinned petrel (*Procellaria aequinoctialis*) on South Georgia: Population size, distribution and global significance. *Polar Biology* 32:655–61.

Martin, J. L., J. C. Thibault, and V. Bretagnolle. 2000. Black rats, island characteristics, and colonial nesting birds in the Mediterranean: Consequences of an ancient introduction. *Conservation Biology* 14:1452–66.

Mason, G., and K. E. Littin. 2003. The humaneness of rodent pest control. *Animal Welfare* 12 (1): 1–37.

Massaro, Melanie, Amanda Starling-Windhof, James V. Briskie, and Thomas E. Martin. 2008. Introduced mammalian predators induce behavioural changes in parental care in an endemic New Zealand bird. *PLoS ONE* 3 (6): e2331. doi:10.1371/journal .pone.0002331(accessed Sept. 30, 2009).

Matias, Rafael, and Paulo Catry. 2008. The diet of feral cats at New Island, Falkland Islands, and impact on breeding seabirds. *Polar Biology* 31 (5): 609–16.

Matisoo-Smith, E., and J. H. Robins. 2004. Origins and dispersals of Pacific peoples: Evidence from mtDNA phylogenies of the Pacific rat. *Proceedings of the National Academy of Sciences* 101 (24): 9167–72.

McCallum, J. 1986. Evidence of predation by kiore upon lizards from the Mokohinau Islands. *New Zealand Journal of Ecology* 9:83–87.

McChesney, Gerard J., and Bernie R. Tershy. 1998. History and status of introduced mammals and impacts to breeding seabirds on the California Channel and Northwestern Baja California islands. *Colonial Waterbirds* 21 (3): 335–47.

McClelland, P., and P. Tyree. 2002. Eradication: The clearance of Campbell Island. *New Zealand Geographic* 58: 86–94.

McClelland, P. J. 2002. Eradication of Pacific rats (*Rattus exulans*) from Whenua Hou Nature Reserve (Codfish Island), Putauhinu and Rarotoka Islands, New Zealand. In *Turning the Tide: The Eradication of Invasive Species*, ed. C. R. Veitch and M. N. Clout, 173–81. Gland, Switzerland and Cambridge, UK: IUCN SSC Invasive Species Specialist Group.

Medway, David G. 2004. The land bird fauna of Stephens Island,

New Zealand in the early 1890s, and the cause of its demise. *Notornis* 51 (4): 201–11.

Meerburg, Bastiaan G., Frans W. A. Brom, and Aize Kijlstra. 2008. The ethics of rodent control. *Pest Management Science* 64 (12): 1205–11.

Merton, D., G. Climo, V. Laboudallon, S. Robert, and C. Mander. 2002. Alien mammal eradication and quarantine on inhabited islands in the Seychelles. In *Turning the Tide: The Eradication of Invasive Species*, ed. C. R. Veitch and M. N. Clout, 182–98. Gland, Switzerland and Cambridge, UK: IUCN SSC Invasive Species Specialist Group.

Merton, Don. 1992. The legacy of "Old Blue." *New Zealand Journal of Ecology* 16 (2): 65–68.

Micol, T., and P. Jouventin. 2002. Eradication of rats and rabbits from Saint-Paul Island, French Southern Territories. In *Turning the Tide: The Eradication of Invasive Species*, ed. C. R. Veitch and M. N. Clout, 199–205. Gland, Switzerland and Cambridge, UK: IUCN SSC Invasive Species Specialist Group.

Milberg, Per, and Tommy Tyrberg. 1993. Naïve birds and noble savages—a review of man-caused prehistoric extinctions of island birds. *Ecography* 16:229–50.

Millar, Heather. 2008. Saving Rat Island. *American Way,* June 1. http://www.americanwaymag.com/rat-island-gregg-howald-alaska-bering-sea (accessed Aug. 6, 2008).

Millener, P. R. 1989. The only flightless passerine; the Stephens Island wren (*Traversia lyalli*: Acanthisittidae). *Notornis* 36 (4): 280–84.

Miskelly, C., and H. Robertson. 2002. Response of forest birds to rat eradication on Kapiti Island (abstract). In *Turning the Tide: The Eradication of Invasive Species*, ed. C. R. Veitch and M. N. Clout, 410. Gland, Switzerland and Cambridge, UK: IUCN SSC Invasive Species Specialist Group.

Miskelly, Colin. 2003. An historical record of bush wren *(Xenicus longipes)* on Kapiti Island. *Notornis* 50:113–14.

Moller, H., and J. L. Craig. 1987. The population ecology of Rattus exulans on Tiritiri Matangi Island, and a model of comparative population dynamics in New Zealand. *New Zealand Journal of Zoology* 14:305–28.

Mooney, H. A., and E. E. Cleland. 2001. The evolutionary impact of invasive species. *Proceedings of the National Academy of Sciences* 98 (10): 5446–51.

Moors, P. J. 1985. Norway rats (*rattus norvegicus*) on the noises and Motukawao islands, Hauraki Gulf, New Zealand. *New Zealand Journal of Ecology* 8:37–54.

Moors, P. J., and I. A. E. Atkinson. 1994. Predation on seabirds by introduced animals, and factors affecting its severity. *International Council for Bird Preservation Technical Publication* 2:667–90.

Morrin, Peter, and Judy Morrin, producers. 2003. *The Battle for Campbell Island*. New Zealand Department of Conservation.

Mouat, Scott, producer. 2010. *The Unnatural History of the Kakapo*. Elwin Productions.

Muir, John. 1965. *The Story of My Boyhood and Youth*. Madison: University of Wisconsin Press.

Mulder, C. P. H., and S. N. Keall. 2001. Burrowing seabirds and reptiles: Impacts on seeds, seedlings and soils in an island forest in New Zealand. *Oecologia* 127:350–60.

Murie, Olaus J. 1959. *Fauna of the Aleutian Islands and Alaska Peninsula*. North American Fauna, no. 61, Washington, D.C.: U.S. Department of Interior, Fish and Wildlife Service.

Myers, Judith H., Daniel Simberloff, Armand M. Kuris, and James R. Carey. 2000. Eradication revisited: Dealing with exotic species. *Trends in Ecology and Evolution* 15 (8): 316–20.

Nettleship, D. N., J. Burger, and M. Gochfeld, eds. 1994. *Seabirds on Islands: Threats, Case Studies and Action Plans*. Cambridge: Birdlife International.

NHNZ TV. 1989. *Island eaten by rats*. http://www.nhnz.tv/nhnz_dvds/product/89.

Nogales, Manuel, Aurelio Martín, Bernie R. Tershy, C. Josh Donlan, Dick Veitch, Néstor Puerta, Bill Wood, and Jesús Alonso. 2004. Invasive rodent eradication on islands. *Conservation Biology* 18 (2): 310–19.

Norman, F. I. 1975. The murine rodents *Rattus rattus, exulans,* and *norvegicus* as avian predators. *Atoll Research Bulletin* 182, 13pp.

Norton, D. A., P. J. deLange, P. J. Garnock-Jones, and D. R. Given. 1997. The role of seabirds and seals in the survival of coastal plants: Lessons from New Zealand Lepidium (Brassicaceae). *Biodiversity and Conservation* 6:765–85.

O'Brien, William. 2006. Exotic invasions, nativism, and ecological restoration: On the persistence of a contentious debate. *Ethics, Place and Environment* 9 (1): 63–77.

O'Donnell, Colin F. J. 1996. Predators and the decline of New Zealand forest birds: An introduction to the hole-nesting bird and predator programme. *New Zealand Journal of Zoology* 23:213–19.

Oglivie, S. C., R. J. Pierce, G. R. G. Wright, L. H. Booth, and C. T. Eason. 1997. Brodifacoum residue analysis in water, soil, invertebrates, and birds after rat eradication on Lady Alice Island. *New Zealand Journal of Ecology* 21 (2): 195–97.

Olson, Storrs L., and Helen F. James. 1989. Fragile outposts. In *Lords of the Air,* ed. J. Page and E. S. Morton, 110–15. Washington, D.C.: Smithsonian Books.

Olson, Storrs L., Leslie Overstreet, and Judith N. Lund. 2007. An Alca-bibliographical study of *The English pilot:* The history of its account of the great auk (Alcidae: *Pinguinus impennis*). *Archives of Natural History* 34 (1): 79–86.

Onion, Amanda. 2005. Studies show rats enjoy tickling. ABC News. http://abcnews.go.com/Technology/story?id=626264&page=2 (accessed July 11, 2010).

Owen, Richard. 1879. On the extinct animals of the colonies of Great Britain. *Popular Science Review* 3:253–73.

Palmer, M., and G. X. Pons. 2001. Predicting rat presence on small islands. *Ecography* 24:121–26.

Panksepp, Jaak, and Jeff Burgdorf. 2003. "Laughing" rats and the evolutionary antecedents of human joy? *Physiology and Behavior* 79:533–47.

Parkes, J. P., N. Macdonald, and G. Leaman. 2002. An attempt to eradicate feral goats from Lord Howe Island. In *Turning the Tide: The Eradication of Invasive Species*, ed. C. R. Veitch and M. N. Clout, 233–39. Gland, Switzerland and Cambridge, UK: IUCN SSC Invasive Species Specialist Group.

Peat, Neville. 2007. Last, loneliest. *Heritage New Zealand*, Spring. http://www.historic.org.nz/en/Publications/HeritageNZ Magazine/HeritageNz2007/HNZ07-LastLoneliest.aspx (accessed May 2, 2010).

Peiser, Benny. 2005. From genocide to ecocide: The rape of Rapa Nui. *Energy and Environment* 16 (3, 4): 513–39.

Pemberton, Mary. 2007. Rats wipe out seabirds on Alaska Island. Associated Press, Nov. 28, 2007. http://www.sfgate.com/cgi-bin/article.cgi?f=/n/a/2007/11/27/national/a123244S87.DTL& feed=rss.news (accessed March 11, 2008).

Pergams, O. R. W., R. C. Lacy, and M. V. Ashley. 2000. Conservation and management of Anacapa deer mice. *Conservation Biology* 14:819–32.

Perrings, Charles, Katharina Dehnen-Schmutz, Julia Touza, and Mark Williamson. 2005. How to manage biological invasions under globalization. *Trends in Ecology and Evolution* 20 (5): 212–15.

Perry, Dan, and Gad Perry. 2008. Improving interactions between animal rights groups and conservation biologists. *Conservation Biology* 22 (1): 27–35.

Petersen, M. R. 1982. Predation on seabirds by red foxes at Shaiak Island, Alaska. *Canadian Field-Naturalist* 96:41–45.

Pierce, R. J. 2002. *Kiore (Rattus exulans) Impact on Breeding Success of*

*Pycroft's Petrels and Little Shearwaters*. Wellington: New Zealand Department of Conservation, DOC Science Internal Series, no. 39.

Pimentel, David, Lori Lach, Rodolfo Zuniga, and Doug Morrison. 2000. Environmental and economic costs of nonindigenous species in the United States. *Bioscience* 50 (1): 53–65.

Pimentel, David, Rodolfo Zuniga, and Doug Morrison. 2005. Update on the environmental and economic costs associated with alien-invasive species in the United States. *Ecological Economics* 52 (3): 273–88.

Pitman, R. L., L. T. Ballance, and C. Bost. 2005. Clipperton Island: Pig sty, rat hole and booby prize. *Marine Ornithology* 33:193–94.

Pitt, William C., and Gary W. Witmer. 2006. Invasive predators: A synthesis of the past, present, and future. In *Internet Center for Wildlife Damage Management, USDA National Wildlife Research Center—Staff Publications*, 264–93. Ft. Collins, CO: National Wildlife Research Center.

Plous, Scott, and Harold A. Herzog. 2000. Poll shows researchers favor lab animal protection. *Science* 290:71.

Poutu, Nick, and Bruce Warburton. 2005. *Effectiveness of the DOC 150, 200 and 250 Traps for Killing Stoats, Ferrets, Norway Rats, Ship Rats, and Hedgehogs*. Wellington: New Zealand Department of Conservation.

Powlesland, Ralph G., Don V. Merton, and John F. Cockrem. 2006. A parrot apart: The natural history of the kakapo (*Strigops habroptilus*), and the context of its conservation management. *Notornis* 53 (1): 3–26.

Pregill, Gregory K., and David W. Steadman. 2004. South Pacific iguanas: Human impacts and a new species. *Journal of Herpetology* 38 (1): 15–21.

Pyšek, Petr, David M. Richardson, and Vojtěch Jarošík. 2006. Who cites who in the invasion zoo: Insights from an analysis of the most highly cited papers in invasion ecology. *Preslia* 78:437–68.

Quammen, David. 1996. *The Song of the Dodo: Island Biogeography in an Age of Extinction.* New York: Scribner.

Quillfeldt, Petra, Ingrid Schenk, Rona A. R. McGill, Ian J. Strange, Juan F. Masello, Anja Gladbach, Verena Roesch, and Robert W. Furness. 2008. Introduced mammals coexist with seabirds at New Island, Falkland Islands: Abundance, habitat preferences, and stable isotope analysis of diet. *Polar Biology* 31 (3): 333–49.

Rapaport, Moshe. 2006. Eden in peril: Impact of humans on Pacific island ecosystems. *Island Studies Journal* 1 (1): 109–24.

Ratcliffe, Norman, Mike Bell, Tara Pelembe, Dave Boyle, Raymond Benjamin, Richard White, Brendan Godley, Jim Stevenson, and Sarah Sanders. 2009. The eradication of feral cats from Ascension Island and its subsequent recolonization by seabirds. *Oryx* 44 (1): 20–29.

Raubenheimer, David, and Stephen J. Simpson. 2006. The challenge of supplementary feeding: Can geometric analysis help save the kakapo? *Notornis* 53 (1): 100–11.

Rauzon, Mark J. 2007. Island restoration: Exploring the past, anticipating the future. *Marine Ornithology* 35:97–107.

Rauzon, Mark J., David Boyle, William T. Everett, and John Gilardi. 2008. The status of the birds of Wake Atoll. *Atoll Research Bulletin* 561, 41pp.

Regehr, Heidi M., Michael S. Rodway, Moira J. F. Lemon, and J. Mark Hipfner. 2007. Recovery of the ancient murrelet *Synthliboramphus antiquus* colony on Langara Island, British Columbia, following eradication of invasive rats. *Marine Ornithology* 35:137–44.

Ricciardi, Anthony. 2007. Are modern biological invasions an unprecedented form of global change? *Conservation Biology* 21 (2): 329–36.

Risbey, Danielle A., Michael C. Calver, Jeff Short, J. Stuart Bradley, and Ian W. Wright. 1999. The impact of cats and foxes on the small vertebrate fauna of Heirisson Prong, Western Australia. II. A field experiment. *Wildlife Research* 27 (3): 223–35.

Roberts, Mere. 1994. A *pakeha* view of the *kiore* rat in New Zealand. In *Science of Pacific Island Peoples: Fauna, flora, food, and medicine*, ed. R. J. Morrison, Paul A. Geraghty, and Linda Crowl, 125–42. Suva, Fiji: University of the South Pacific.

Robertson, Bruce C. 2006. The role of genetics in kakapo recovery. *Notornis* 53 (1): 173–83.

Robertson, H. A., R. M. Colbourne, and F. Nieuwland. 1993. Survival of little spotted kiwi and other forest birds exposed to brodifacoum rat poison on Red Mercury Island. *Notornis* 40:253–62.

Rodda, G. H., T. H. Fritts, E. W. Campbell III, K. Dean-Bradley, G. Perry, and C. P. Qualls. 2002. Practical concerns in the eradication of island snakes. In *Turning the Tide: The Eradication of Invasive Species*, ed. C. R. Veitch and M. N. Clout, 260–65. Gland, Switzerland and Cambridge, UK: IUCN SSC Invasive Species Specialist Group.

Rodionov, S. N., J. E. Overland, and N. A. Bond. 2005. Spatial and temporal variability of the Aleutian climate. *Fisheries Oceanography* 14 (supp. 1): 3–21.

Rolston, Holmes III. 2003. Environmental ethics. In *The Blackwell Companion to Philosophy*, 2nd ed., ed. Nicholas Bunnin and E. P. Tsui-James, 517–30. Malden, MA: Blackwell Publishers.

Rooney, Katie. 2008. The top ten everything of 2008. 10: $150,000 for "Rat Island." *Time*, Nov. 3. http://www.time.com/time/specials/packages/article/0,28804,1855948_1863903_1863888,00 .html (accessed Nov. 17, 2010).

Rosen, Yereth. 2007. Biologists aim to wipe out "Rat Island." Reuters, Oct. 2. http://www.reuters.com/article/idUSN01297168 20071002 (accessed Nov. 17, 2010).

————. 2009. Grand quest to rid island of rats. *Christian Science Monitor*, March 12. http://www.csmonitor.com/The-Culture/2009/0312/a-grand-quest-to-rid-an-island-of-rats (accessed Nov. 17, 2010).

Rowsell, H. C., J. Ritcey, and F. Cox. 1979. Assessment of humaneness of vertebrate pesticides. In *Proceedings of the Canadian Association for Laboratory Animal Science 1978–1979*, 159–249.

Rudman, B. 2003. Latest twist in rat saga right up kiore's alley. *New Zealand Herald*, May 9.

Ruffino, L., K. Bourgeois, E. Vidal, J. Icard, F. Torre, and J. Legrand. 2008. Introduced predators and cavity-nesting seabirds: Unexpected low level of interaction at breeding sites. *Canadian Journal of Zoology* 86:1068–73.

Russell, J. C., M. N. Clout, and B. H. McArdle. 2004. Island biogeography and the species richness of introduced mammals on New Zealand offshore islands. *Journal of Biogeography* 31:653–64.

Russell, J. C., D. R. Towns, and M. N. Clout. 2008. Review of rat invasion biology: Implications for island biosecurity. *Science for Conservation* 286. New Zealand Department of Conservation, Wellington. 53pp.

Russell, James. 2006. Overlooked damages. *Conservation* 8 (3): 46.

Russell, James C., Brent M. Beaven, Jamie W. B. MacKay, David R. Towns, and Mick N. Clout. 2008. Testing island biosecurity systems for invasive rats. *Wildlife Research* 35:215–21.

Russell, James C., and Matthieu Le Corre. 2009. Introduced mammal impacts on seabirds in the Îles Éparses, Western Indian Ocean. *Marine Ornithology* 37:121–29.

Russell, James C., Vincent Lecomte, Yves Dumont, and Matthieu Le Corre. 2009. Intraguild predation and mesopredator release effect on long-lived prey. *Ecological Modelling* 220:1098–1104.

Russell, James C., Jamie W. B. Mackay, and Jawad Abdelkrim. 2009. Insular pest control within a metapopulation context. *Biological Conservation* 142:1404–10.

Russell, James C., David R. Towns, Sandra H. Anderson, and Mick N. Clout. 2005. Intercepting the first rat shore. *Nature* 437:1107.

Russell, James Charles. 2004. Invading the Pacific: Biological and cultural dimensions of invasive species in the Pacific region. *Graduate Journal of Asia-Pacific Studies* 2 (2): 77–94.

Safina, Carl. 2002. *Eye of the Albatross: Visions of Hope and Survival.* New York: Henry Holt.

Sagoff, Mark. 2005. Do non-native species threaten the natural environment? *Journal of Agricultural and Environmental Ethics* 18:215–36.

Samways, M. J., S. Taylor, and W. Tarboton. 2005. Extinction reprieve following alien removal. *Conservation Biology* 19:1329–30.

Sanders, Mark D., and Richard F. Maloney. 2002. Causes of mortality at nests of ground-nesting birds in the Upper Waitaki Basin, South Island, New Zealand: A 5-year video study. *Biological Conservation* 106:225–36.

Sappenfield, M. 2002. Off the California coast, it's alien rats versus native birds. *Christian Science Monitor*, Oct. 31.

Saunders, A., and D. A. Norton. 2001. Ecological restoration at Mainland Islands in New Zealand. *Biological Conservation* 99: 109–19.

Sax, Dov F., and Steven D. Gaines. 2008. Species invasions and extinction: The future of native biodiversity on islands. *Proceedings of the National Academy of Sciences* 105 (supp. 1): 11490–97.

Sax, Dov F., Steven D. Gaines, and James H. Brown. 2002. Species invasions exceed extinctions on islands worldwide: A comparative study of plants and birds. *American Naturalist* 160 (6): 766–83.

Schwartz, John. 2005. Rat on the run turns out to be marathon swimmer. *New York Times*, Oct. 25.

Seddon, P. J., and R. F. Maloney. 2003. *Campbell Island Teal Re-Introduction Plan.* Wellington: New Zealand Department of Conservation, DOC Science Internal Series, no. 154.

Sekercioglu, Cagan H. 2006. Increasing awareness of avian ecological function. *Trends in Ecology and Evolution* 21(8): 464–71.

Şekercioğlu, Çağan H., Gretchen C. Daily, and Paul R. Ehrlich. 2004. Ecosystem consequences of bird declines. *Proceedings of the National Academy of Sciences* 101 (52): 18042–47.

Sessions, Laura. 2003. Date with extinction. *Natural History* 112 (3): 52–58.

Shaffer, Scott A., Yann Tremblay, Henri Weimerskirch, Darren Scott, David R. Thompson, Paul M. Sagar, Henrik Moller et al. 2006. Migratory shearwaters integrate oceanic resources across the Pacific Ocean in an endless summer. *Proceedings of the National Academy of Sciences* 103 (34): 12799–802.

Shah, N. J. 2001. Eradication of alien predators in the Seychelles: An example of conservation action on tropical islands. *Biodiversity and Conservation* 10:1219–20.

Sharp, Trudy, and Glen Saunders. 2008. *A Model for Assessing the Relative Humaneness of Pest Animal Control Methods.* Canberra: Australian Government Department of Agriculture, Fisheries and Forestry.

Sherley, Greg, ed. 2000. *Invasive Species in the Pacific: A Technical Review and Draft Regional Strategy.* Samoa: South Pacific Regional Environment Programme.

Siebert, Charles. 2010. The animal-cruelty syndrome. *New York Times Magazine,* June 7. http://www.nytimes.com/2010/06/13/magazine/13dogfighting-t.html?WT.mc_id=MG-SM-E-FL-SM-LIN-TSO-061110-NYT-NA&WT.mc_ev=click&pagewanted=all (accessed June 23, 2010).

Simberloff, Daniel. 2000. Extinction-proneness of island species: Causes and management implications. *Raffles Bulletin of Zoology* 48:1–9.

———. 2001. Eradication of island invasives: Practical actions and results achieved. *Trends in Ecology and Evolution* 16:273–74.

———. 2002. Today Tiritiri Matangi, tomorrow the world! Are we aiming too low in invasives control? In *Turning the Tide: The Eradication of Invasive Species*, ed. C. R. Veitch and M. N. Clout, 4–12. Gland, Switzerland and Cambridge, UK: IUCN SSC Invasive Species Specialist Group.

———. 2003. Confronting introduced species: A form of xenophobia? *Biological Invasions* 5:179–92.

———. 2006. Invasional meltdown 6 years later: Important phenomenon, unfortunate metaphor, or both? *Ecology Letters* 9 (8): 912–19.

———. 2006. Rejoinder to Simberloff (2006): Don't calculate effect sizes; study ecological effects. *Ecology Letters* 9 (8): 921–22.

———. 2008. Successes, failures, and challenges in protecting biodiversity: DOC and the next 20 years. In *Proceedings of the Conservvision Conference*, University of Waikato, July 2–4. Hamilton, NZ: University of Waikato.

Simberloff, Daniel, and Betsy Von Holle. 1999. Positive interactions of nonindigenous species: Invasional meltdown? *Biological Invasions* 1:21–32.

Singer, Peter. 1990. *Animal Liberation*. New York: Avon Books.

Smith, D. G., E. K. Shiinoki, and E. A. VanderWerf. 2006. Recovery of native species following rat eradication on Mokoli'i Island, O'ahu, Hawai'i. *Pacific Science* 60 (2): 299–303.

Smith, Roff. 2008. Beyond the blue horizon: How ancient voyagers settled the far-flung islands of the Pacific. *National Geographic*, March. http://ngm.nationalgeographic.com/2008/03/people-pacific/smith-text (accessed Oct. 16, 2009).

Soule, Michael E. 1990. The onslaught of alien species, and other challenges in the coming decades. *Conservation Biology* 4:233–39.

South Georgia Heritage Trust. 2009. Environmental impact assessment for the eradication of rodents from the island of South Georgia, 82pp.

———. 2010. Operational plan for the eradication of rodents from South Georgia: Phase 1, 54pp.

Southwell, Ben, producer. 2010. *Last Chance to See: Kakapo*. BBC Wales/West Park Pictures.

Stapp, P. 2002. Stable isotopes reveal evidence of predation by ship rats on seabirds on the Shiant Islands, Scotland. *Journal of Applied Ecology* 39:831–40.

Steadman, D. W. 1995. Prehistoric extinctions of Pacific island birds: Biodiversity meets zooarchaeology. *Science* 267:1123–31.

Steadman, D. W., and P. S. Martin. 2003. The late Quaternary extinction and future resurrection of birds on Pacific islands. *Earth-Science Reviews* 61:133–47.

Steadman, David W. 1995. Extinction of birds on tropical Pacific islands. In *Late Quaternary Environments and Deep History: A Tribute to Paul S. Martin*, ed. D. W. Steadman and J. I. Mead, 33–49. Hot Springs, SD: The Mammoth Site.

————. 1996. Extinctions of Polynesian birds: Reciprocal impacts of birds and people. In *Historical Ecology in the Pacific Islands*, ed. P. V. Kirch and T. L. Hunt, 51–79. New Haven, CT: Yale University Press.

Steadman, David W., and Storrs L. Olson. 1985. Bird remains from an archaeological site on Henderson Island, South Pacific: Man-caused extinctions on an "uninhabited" island. *Proceedings of the National Academy of Sciences* 82 (18): 6191–95.

Steadman, David W., Gregory K. Pregill, and David V. Burley. 2002. Rapid prehistoric extinction of iguanas and birds in Polynesia. *Proceedings of the National Academy of Sciences* 99 (6): 3673–77.

Steadman, David W., J. Peter White, and Jim Allen. 1999. Prehistoric birds from New Ireland, Papua New Guinea: Extinctions on a large Melanesian island. *Proceedings of the National Academy of Sciences* 96 (5): 2563–68.

Steller, Georg Wilhelm. 1988. *Journal of a Voyage with Bering, 1741–1742*. Ed. O. W. Frost and M. A. Engel. Stanford, CA: Stanford University Press.

Sullivan, Robert. 2004. *Rats: Observations on the History and Habitat of the City's Most Unwanted Inhabitants*. New York: Bloomsbury USA.

Taborsky, M. 1988. Kiwis and dog predation: Observations in Waitangi State Forest. *Notornis* 35:197–202.

Tanji, Melissa. 2008. Birds vital to ecosystem being killed by pets and feral animals—officials. *Maui News*, December 5. http://www.mauinews.com/page/content.detail/id/512046.html?nav=10 (accessed Aug. 11, 2009).

Taylor, R. H. 1975. What limits kiore (*Rattus exulans*) distribution in New Zealand? *New Zealand Journal of Zoology* 2:473–77.

———. 1979. How the Macquarie Island parakeet became extinct. *New Zealand Journal of Ecology* 2:42–45.

Taylor, R. H., and B. W. Thomas. 1993. Rats eradicated from rugged Breaksea Island (170 Ha), Fiordland, New Zealand. *Biological Conservation* 65:191–98.

Taylor, Rowland H., Gary W. Kaiser, and Mark C. Drever. 2000. Eradication of Norway rats for recovery of seabird habitat on Langara Island, British Columbia. *Restoration Ecology* 8:151–60.

Temple, Stanley A. 1990. The nasty necessity: Eradicating exotics. *Conservation Biology* 4 (2): 113–15.

Tershy, B. R., C. J. Donlan, B. Keitt, D. Croll, J. A. Sanchez, B. Wood, M. A. Hermosillo, and G. Howald. 2002. Island conservation in Northwest Mexico: A conservation model integrating research, education and exotic mammal eradication. In *Turning the Tide: The Eradication of Invasive Species*, ed. C. R. Veitch and M. N. Clout, 293–300. Gland, Switzerland and Cambridge, UK: IUCN SSC Invasive Species Specialist Group.

Thomas, B. W., and R. H. Taylor. 2002. A history of ground-based rodent eradication techniques developed in New Zealand, 1959–1993. In *Turning the Tide: The Eradication of Invasive Species*, ed. C. R. Veitch and M. N. Clout, 301–10. Gland, Switzerland and Cambridge, UK: IUCN SSC Invasive Species Specialist Group.

Thorsen, M., R. Shorten, R. Lucking, and V. Lucking. 2000. Norway rats (*Rattus norvegicus*) on Frégate Island, Seychelles: The invasion; subsequent eradication attempts and implications for the island's fauna. *Biological Conservation* 96:133–38.

Tillotson, Mary. 2010. Paradise threatened by Rattus rattus. *U.S. Fish and Wildlife Service Refuge Update* 7 (3): 16.

Tipa, Rob. 2006. The kakapo in Maori lore. *Notornis.* 53:193–94.

Tompkins, Jennifer. 2000. Eradication of *Rattus norvegicus* from seabird habitat in Canada. *Restoration and Reclamation Review* 6 (6): 1–6. http://www.ecoinfo.org/env_ind/region/seabird/seabird .htm (accessed July 9, 2009).

Towns, D. R. 1994. The role of ecological restoration in conservation of Whitaker's skink (Cyclodina whitakeri), a rare New Zealand lizard (Lacertilia: Scincidae). *New Zealand Journal of Zoology* 21:457–71.

———. 2002. Interactions between geckos, honeydew scale insects and host plants revealed on islands in northern New Zealand, following eradication of introduced rats and rabbits. In *Turning the Tide: The Eradication of Invasive Species*, ed. C. R. Veitch and M. N. Clout, 329–35. Gland, Switzerland and Cambridge, UK: IUCN SSC Invasive Species Specialist Group.

———. 2002. Korapuki Island as a case study for restoration of insular ecosystems in New Zealand. *Journal of Biogeography* 29:593–607.

Towns, D. R., D. Simberloff, and I. A. E. Atkinson. 1997. Restoration of New Zealand islands: Redressing the effects of introduced species. *Pacific Conservation Biology* 3:99–124.

Towns, David R., I. A. E. Atkinson, and C. H. Daugherty. 2006. Have the harmful effects of introduced rats on islands been exaggerated? *Biological Invasions* 8:863–91.

Towns, David R., and W. J. Ballantine. 1993. Conservation and restoration of New Zealand island ecosystems. *Trends in Ecology and Evolution* 8 (12): 452–57.

Towns, David R., and Keith G. Broome. 2003. From small Maria to massive Campbell: Forty years of rat eradications from New Zealand islands. *New Zealand Journal of Zoology* 30:377–98.

Towns, David R., and Charles H. Daugherty. 1994. Patterns of range contractions and extinctions in the New Zealand herpetofauna following human colonisation. *New Zealand Journal of Zoology* 21:325–39.

Towns, David R., C. H. Daugherty, and A. Cree. 2001. Raising the prospects for a forgotten fauna: A review of 10 years of conservation for New Zealand reptiles. *Biological Conservation* 99:3–16.

Towns, David R., and S. M. Ferreira. 2001. Conservation of New Zealand's lizards (Lacertilia: Scincidae) by translocation of small populations. *Biological Conservation* 98:211–22.

Towns, David R., Richard Parrish, Claudine L. Tyrrell, Graham T. Ussher, Alison Cree, Donald G. Newman, A. (Tony) H. Whitaker, and Ian Westbrooke. 2007. Responses of tuatara (*Sphenodon punctatus*) to removal of introduced Pacific rats from islands. *Conservation Biology* 21 (4): 1021–31.

Towns, David R., G. Richard Parrish, and Ian Westbrooke. 2003. Inferring vulnerability to introduced predators without experimental demonstration: Case study of Suter's skink in New Zealand. *Conservation Biology* 17 (5): 1361–71.

Trevino, Heather S., Amy L. Skibiel, Tim J. Karels, and F. Stephen Dobson. 2007. Threats to avifauna on oceanic islands. *Conservation Biology* 21 (1): 125–32.

Tyrrell, C. L., A. Cree, and D. R. Towns. 2000. *Variation in Reproduction and Condition of Northern Tuatara (Sphenodon punctatus punctatus) in the Presence and Absence of Kiore*. Science for Conservation, no. 153. Wellington: New Zealand Department of Conservation.

U.S. Fish and Wildlife Service. 2007. *Restoring Wildlife Habitat on Rat Island, Alaska Maritime National Wildlife Refuge, Aleutian Islands Unit, Environmental Assessment*. Homer, AK: Aleutian Island National Wildlife Refuge.

————. 2009. Reports from Rat Island reflect successes and concerns. Press release, June 11.

————. 2009. No rats found, lab results on six bird tests received. Press release, July 1.

U.S. National Park Service. 2003. Rat interactions with native wildlife. http://www.nps.gov/chis/naturalresources (accessed Jan. 27, 2004).

Ussher, G. T. 1999. Tuatara (*Sphenodon punctatus*) feeding ecology in the presence of kiore (*Rattus exulans*). *New Zealand Journal of Zoology* 26:117–25.

Van Driesche, J., and R. Van Driesche. 2004. *Nature Out of Place: Biological Invasions in the Global Age*. Washington, D.C.: Island Press.

Van Rensburg, P. J. J., and M. N. Bester. 1988. The effect of cat *Felis catus* predation on three breeding Procellariidae species on Marion Island. *South African Journal of Zoology* 23 (4): 301–05.

Vantassel, Stephen. 2008. Ethics of wildlife control in humanized landscapes: A response. In *Proceedings of the Twenty-third Vertebrate Pest Conference*, ed. R. M. Timm and M. B. Madon, 294–300. Davis: University of California.

Veitch, C. R. 2001. The eradication of feral cats (*Felis catus*) from Little Barrier Island, New Zealand. *New Zealand Journal of Zoology* 28:1–12.

Veitch, C. R., and M. N. Clout, ed. 2002. *Turning the Tide: The Eradication of Invasive Species*. Proceedings of the International Conference on Eradication of Island Invasives. Gland, Switzerland and Cambridge, UK: IUCN SSC Invasive Species Specialist Group.

Veltman, C. 1996. Investigating causes of population decline in New Zealand plants and animals: Introduction to a symposium. *New Zealand Journal of Ecology* 20:1–5.

Wade, Nicholas. 2007. Study traces cat's ancestry to Middle East. *New York Times*, June 29. http://www.nytimes.com/2007/06/29/science/29cat.html?_r=1&em&ex=1183348800&en=46920e3fe2f7c649&ei=5087%0A (accessed Dec. 22, 2008).

Wallace, George, and Joni Ellis. 2003. *Issue Assessment: Impacts of Feral and Free-ranging Domestic Cats on Wildlife in Florida*. Unpublished report, Florida Fish and Wildlife Conservation Commission.

Walsh, Julie, Kerry-Jayne Wilson, and Graeme P. Elliott. 2006. Seasonal changes in home range size and habitat selection by kakapo (*Strigops habroptilus*) on Maud Island. *Notornis* 53 (1): 143–49.

Wanless, Ross M., Andrea Angel, Richard J. Cuthbert, Geoff M. Hilton, and Peter G. Ryan. 2007. Can predation by invasive mice drive seabird extinctions? *Biology Letters* 3:241–44.

Warburton, B., and B. G. Norton. 2009. Towards a knowledge-based ethic for lethal control of nuisance wildlife. *Journal of Wildlife Management* 73 (1): 158–64.

Warburton, B., Nick Poutu, and Ian Domigan. 2002. *Effectiveness of the Victor Snapback Trap for Killing Stoats*. DOC Science Internal Series, no. 83. Wellington: New Zealand Department of Conservation.

Warburton, Bruce. 2008. Invasive species. In *Encyclopedia of Environmental Ethics and Philosophy*, ed. J. Baird Callicott and Robert Frodeman, 531–33. Detroit: Macmillan Reference USA.

Warne, Kennedy. 2002. Hotspots: New Zealand. *National Geographic*, October, 75–101.

Waymer, Jim. 2007. Bird extinctions may quicken: Loss of habitat, warming, cats are all factors. *Florida Today*, Jan. 23. http://www.floridatoday.com/apps/pbcs.dll/article?AID=/20070123/NEWS01/701230332/1006&template=printart (accessed Feb. 13, 2008).

Whitaker, A. H. 1978. The effects of rodents on reptiles and amphibians. In *The Ecology and Control of Rodents in New Zealand Nature Reserves*, ed. P. R. Dingwall, I. A. E. Atkinson, and C. Hay, 75–86. Wellington: Department of Lands and Survey Information Series 4.

White, Kevin, producer. 2003. *Restoring Balance: Removing the Black Rat from Anacapa Island*. Filmmakers Collaborative, San Francisco.

White, Piran C. L., Adriana E. S. Ford, Mick N. Clout, Richard M. Engeman, Sugoto Roy, and Glen Saunders. 2008. Alien invasive vertebrates in ecosystems: Pattern, process and the social dimension. *Wildlife Research* 35:171–79.

White, Taylor. 1894. Remarks on the rats of New Zealand. *Transactions and Proceedings of the Royal Society of New Zealand 1868–1961* 27:240–61.

Whitworth, D. L., H. R. Carter, J. S. Koepke, and F. Gress. 2008. *Nest Monitoring of Xantus's Murrelets at Anacapa Island, California: 2007 Annual Report*. Unpublished report, California Institute of Environmental Studies, Davis, 33 pp.

Wilcove, D. S., D. Rothstein, J. Dubow, A. Phillips, and E. Losos. 1998. Quantifying threats to imperiled species in the United States. *BioScience* 48:607–15.

Wilcox, Chris, and C. Josh Donlan. 2007. Compensatory mitigation as a solution to fisheries bycatch–biodiversity conservation conflicts. *Frontiers in Ecology and Environment* 5 (6): 325–31.

Williams, G. R. 1956. The kakapo (*Strigops habroptilus*, Gray): A review and re-appraisal of a near-extinct species. *Notornis* 7 (2): 29–56.

Williams, Ted. 2009. Felines fatales. *Audubon*, September/October. http://audubonmagazine.org/incite/incite0909.html (accessed Sept. 29, 2010).

Wilmshurst, J. M., and T. F. G. Higham. 2004. Using rat-gnawed seeds to independently date the arrival of Pacific rats and humans in New Zealand. *Holocene* 14:801–06.

Wilmshurst, Janet M., Atholl J. Anderson, Thomas F. G. Higham, and Trevor H. Worthy. 2008. Dating the late prehistoric dispersal of Polynesians to New Zealand using the commensal Pacific rat. *Proceedings of the National Academy of Sciences* 105 (22): 7676–80.

Wilson, Deborah J., A. D. Grant, and N. Parker. 2006. Diet of kakapo in breeding and non-breeding years on Codfish Island (Whenua Hou) and Stewart Island. *Notornis* 53 (1): 80–89.

Wood, B., B. R. Tershy, M. A. Hermosillo, C. J. Donlan, J. A. Sanchez, B. S. Keitt, D. A. Croll, G. R. Howald, and N. Biavaschi. 2002. Removing cats from islands in north-west Mexico. In *Turning the Tide: The Eradication of Invasive Species*, ed. C. R. Veitch and M. N. Clout, 374–80. Gland, Switzerland and Cambridge, UK: IUCN SSC Invasive Species Specialist Group.

Wood, Jamie R. 2006. Subfossil kakapo (*Strigops habroptilus*) remains from near Gibraltar Rock, Cromwell Gorge, Central Otago, New Zealand. *Notornis* 53:191–93.

Worthy, T. H., and R. N. Holdaway. 2002. *The Lost World of the Moa: Prehistoric Life of New Zealand*. Christchurch, New Zealand: Canterbury University Press.

Yabe, Tatsuo, Takuma Hashimoto, Masaaki Takiguchi, Masanari Aoki, and Kazuto Kawakami. 2009. Seabirds in the stomach contents of black rats *Rattus rattus* on Higashijima, the Ogasawara (Bonin) Islands, Japan. *Marine Ornithology* 37:293–95.

Zelle, Carolyn, producer. 2003. *Journey of the Tiĝlax̂*. Odyssey Productions.

# INDEX

# A NOTE ON THE AUTHOR

WILLIAM STOLZENBURG has written hundreds of magazine articles about the science and spirit of saving wild creatures. A 2010 Alicia Patterson Journalism Fellow, he is the author of the book *Where the Wild Things Were* and the screenwriter of the documentary *Lords of Nature: Life in a Land of Great Predators.* He lives in Shepherdstown, West Virginia, and Fairfax, Virginia. Visit his Web site at www.williamstolzenburg.com.